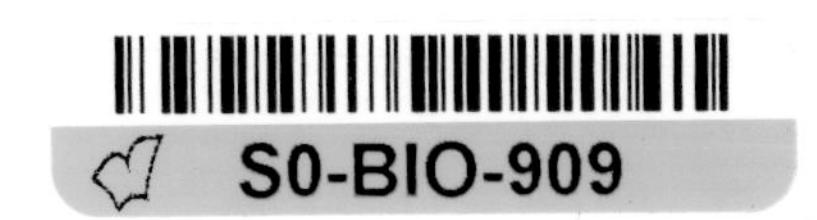

THE LANGUAGE OF FIRST-ORDER LOGIC

CSLI Lecture Notes Number 23

THE LANGUAGE OF FIRST-ORDER LOGIC

Third Edition, Revised & Expanded

Including the Macintosh version of

TARSKI'S WORLD 4.0

Jon Barwise & John Etchemendy

CSLI PUBLICATIONS
Center for the Study of Language and Information
STANFORD
CALIFORNIA

CSLI was founded early in 1983 by researchers from Stanford University, SRI International, and Xerox PARC to further research and development of integrated theories of language, information, and computation. CSLI headquarters and the publication offices are located at the Stanford site.

CSLI/SRI International
333 Ravenswood Avenue
Menlo Park, CA 94025

CSLI/Stanford
Ventura Hall
Stanford, CA 94305

CSLI/Xerox PARC
3333 Coyote Hill Road
Palo Alto, CA 94304

Copyright ©1993
Center for the Study of Language and Information
Leland Stanford Junior University

Printed in the United States

01 00 99 98 97 96 95 94 93 5 4 3 2 1

Library of Congress Cataloging-in-Publication Data

Barwise, Jon

The language of first-order logic : including the Macintosh program Tarski's world 4.0 / Jon Barwise and John Etchemendy. – 3rd ed., rev. and expanded.

p. cm. – (CSLI lecture notes ; no. 23)

Includes indexes.

ISBN 0–937073–99–7 (pbk.)

1. First-order logic. 2. Tarski's world (Computer program). I. Etchemendy, John, 1952– . II. Tarski's world (Computer program). 1993. III. Title. IV. Series.

BC128.B37 1993

160—dc20 93-419

CIP

Please note that this book and the diskette accompanying it may not be returned after the pouch containing the diskette has been unsealed. The diskette, Tarski's World 4.0, is protected by copyright and may not be duplicated, except for purposes of archival back up. All inquiries concerning the diskette should be directed to the Publications Office, CSLI, Ventura Hall, Stanford University, Stanford, CA 94305; phone: (415)723–1712 or (415)723–1839. The program has been duplicated on the finest quality diskettes, verified before shipment. If a diskette is damaged, please send it to the above address for replacement. Please note also that the distributor of this book and program, the University of Chicago Press, is not equipped to handle inquiries about this package, nor to replace defective diskettes.

Microsoft® is a registered trademark of Microsoft Corporation.
Windows™ is a trademark licensed to Microsoft Corporation.
IBM® is a registered trademark of IBM Corporation.
Apple® is a registered trademark of Apple Computer, Incorporated.
Macintosh™ is a trademark licensed to Apple Computer, Incorporated.
NeXT™ is a trademark licensed to NeXT Computer, Incorporated.

To the people who made Tarski's World a reality:
Rick Wong, Rolf van Widenfelt, Steve Loving,
Alan Bush, Mike Lenz, Eric Ly, Pete Murray,
Dan Fish, Kalpana Bharadwaj,
and Christopher Fuselier.

Preface

This book was designed for use in a first course in logic. Our primary intended audience is students interested in the symbolic sciences (artificial intelligence, computer science, linguistics, philosophy, and mathematics), but we have tried to make it interesting and accessible to all. It has no prerequisites beyond high school algebra, and is suitable for undergraduates with very different backgrounds and mathematical abilities, and with very different career goals in mind.

The book covers more than enough material for a semester course, but it is organized so students will get a good introduction by getting through Part II, or even just through Chapter 6. This can be done in ten weeks without pushing too hard. Chapter 6 is really the high point of the book, and is the chapter that takes the longest to work through. In a semester, you can cover Parts I–III, or the first six or seven chapters and Part IV. Part III focuses more on mathematical topics (set theory and induction), while Part IV concentrates on topics from logic and computer science, including resolution and unification.

Pedagogically, the book is based on our experience that students learn logic best by getting involved with the material through problems. For this reason, there are hundreds of exercises and problems. In our own courses, we encourage students to work on problems in small groups of two to four students. We ask them to write up and turn in their problem sets individually, however. We find that in working on the problems together, students discover (and usually sort out) their own misconceptions much more efficiently than when working alone. By writing up their own solutions, they are less likely to trick themselves about whether they understand the solutions arrived at by their group.

The book is designed for use with Tarski's World, a program available on Macintosh and NeXT computers, and IBM PC's equipped with Microsoft Windows. The Macintosh and IBM versions of the program come packaged in separate editions of this book. To receive a copy of the NeXT version, send your original disk from the back of this book to CSLI Publications, Ventura Hall, Stanford CA 94305-4115. Include a self-addressed, stamped envelope and $2.00 to cover the cost of the NeXT diskette.

Tarski's World is an essential part of the course. Most of the problems in Parts I, II, and IV use the program. The two of us do not use a computer in the classroom itself, though this is mainly for logistic reasons. Rather, we have our students work on computers scattered around campus, outside of class. This has not been a problem. If your class does have access to a classroom with a computer and projector, however, it would no doubt allow for more innovative lectures and for useful classroom discussion of Tarski's World problems.

We distinguish between "exercises" and "problems." Problems result in work that can be handed in and graded, while exercises do not. This does not mean that students should skip the exercises. Many of them are important for understanding the material. We think of it as similar to a college sport. You can't learn how to play the game by sitting in lectures. You need to practice (the exercises) before you can get out and play the game (the problems).

Problems and exercises that make use of Tarski's World are indicated with a diamond ($\diamond$) subscript, which is meant to call to mind the blocks of Tarski's World. We also use a series of stars ($\star$, $\star\star$, $\star\star\star$) as a rough indication of the difficulty of the problem. So, for example, **Problem**$^{\star}_{\diamond}$ indicates a problem that is not routine, and that uses Tarski's World.

Note to the instructor

An instructor's manual and instructor's disk are available to accompany this book. The instructor's manual, written by Ruth Eberle, provides a wealth of material to make grading exercises and teaching from the text easier. The instructor's disk allows you to grade most of the Tarski's World problems automatically, directly off the student's disks.

To receive the manual and disk, send an informal but rigorous proof that you have adopted the text to: CSLI Publications, Ventura Hall, Stanford University, Stanford, CA 94305-4115. Be sure your proof makes clear which version of Tarski's World (i.e., Macintosh, IBM, or NeXT) you and your students are using.

Although it simplifies things somewhat if all of your students use the same platform, you can also let them choose which computer they prefer. The Macintosh version of Tarski's World 4.0 can grade both Macintosh and IBM disks (and NeXT disks using IBM or Macintosh formats). To grade these disks simultaneously, you will need access to a Macintosh running System 7 and equipped with Macintosh PC

Exchange (available from Apple). Complete instructions are included on the Macintosh instructor's disk.

IBM-compatible computers cannot read Macintosh disks, and so of course cannot grade them. However students using the Macintosh version of Tarski's World 4.0 can transfer their homework files onto IBM disks (using Apple File Exchange or Macintosh PC Exchange), and turn in these disks to be graded. The files themselves are completely compatible across platforms.

If you decide to use a single platform in your class, we would personally recommend using the Macintosh. We have found that students without computer experience effortlessly learn to use Tarski's World on the Macintosh, and the Macintosh, unlike the NeXT, is widely available.

Acknowledgements

Our primary debt of gratitude goes to the programmers who made Tarski's World such a wonderful program. The original version was written by Rick Wong and Rolf van Widenfelt, under the guidance of Steve Loving. Developing the original program took over three years from our initial conception to the first released version. Rick, Rolf, and Steve's work on the program was supported by the Faculty Author Development Program at Stanford University. The current Macintosh version of Tarski's World incorporates numerous improvements to the original program, and itself took many years to complete. The programming was carried out by Mike Lenz, except for the grading facility, which was written by Alan Bush.

The growing success of the package over the past few years convinced us to develop versions for other machines. This development has been overseen by the second author. The NeXT and Windows versions share a computational engine, but required completely distinct user interface code. The common engine was written by Eric Ly, Pete Murray, and Dan Fish. Eric wrote the NeXT interface single-handedly in a matter of months, a great tribute both to Eric and to the NeXTstep development platform. The Windows interface was much more difficult, due to the complexites of the Windows environment. The interface was written by Christopher Fuselier and Kalpana Bharadwaj, building on earlier code written by Pete and Dan. Except for the original version, all work was funded by the Center for the Study of Language and Information.

People who have been involved in software development will realize what a tremendous debt we owe to all ten of our programmers. The creativity, orginality, and pure hard work that goes into developing an ambitious piece of software is too often overlooked or undervalued. Finally, but perhaps most important, we have gotten a lot of personal pleasure out of working with all of our Tarski's World programmers over the past decade, and take our hats off to them for a job well done.

Finally, we would like to thank all those students and colleagues who have made important suggestions about the textbook, especially Carl Ginet, Daniel Quesada, Bill Hanson, Ernest Adams, Tom Burke, Pete Woodruff, Carol Cleland, Tom Wasow, Eric Hammer, Ruth Eberle, and Jon Russell Barwise. We trust that there will be further improvements suggested by future users, and invite written comments at either of the following addresses:

Jon Barwise
Department of Philosophy
Sycamore Hall 120
Indiana University
Bloomington, IN 47405

E-mail: barwise@phil.indiana.edu

John Etchemendy
C.S.L.I.
Ventura Hall
Stanford University
Stanford, CA 94305

E-mail: etch@csli.stanford.edu

Contents

1

Introduction

1.1 The special role of logic in rational inquiry

What do the fields of astronomy, economics, law, mathematics, physics, and sociology have in common? Not much in the way of subject matter, that's for sure. And not all that much in the way of methodology. What they do have in common, though, with each other and with many other fields, is their dependence on a certain standard of rationality. In each of these fields, it is assumed that the participants can differentiate between rational argumentation based on accepted principles, and wild speculation or total nonsequiturs. In other words, these fields all presuppose an underlying acceptance of basic principles of logic.

For that matter, *all* rational inquiry depends on logic, on the ability of people to reason correctly most of the time. And, when they fail to reason correctly, on the ability of others to point out the gaps in their reasoning. While people may not all agree on a whole lot, they do seem to be able to agree on what constitutes a legitimate conclusion from given premises. Acceptance of these commonly held principles of rationality is what differentiates rational inquiry from other forms of human activity.

Just what are the principles of rationality that underwrite these disciplines? And what are the techniques by which we can distinguish valid argumentation from invalid argumentation? More basically, what is it that *makes* one claim follow from accepted premises, while some other claim does not?

Many answers to these questions have been explored. One suggestion that still has its adherents is that the laws of logic are a matter of convention. If this is so, we could presumably decide to change the conventions, and so adopt different principles of logic, the way we can decide which side of the road we drive on. But there is an overwhelming intuition that the laws of logic are somehow more irrefutable than the laws of the land, or even the laws of physics.

The importance of logic has been recognized since antiquity. After all, no science can be any more certain than its weakest link. If there is something arbitrary about logic, then the same must hold of all rational inquiry. Thus it becomes crucial to understand just what the laws of logic are, and even more important, *why* they are laws of logic. These are the questions that one takes up when one studies logic itself. To study logic is to use the methods of rational inquiry on rationality itself.

Over the past century the study of logic has undergone rapid and important advances. Spurred on by logical problems in that most deductive of disciplines, mathematics, it developed into a discipline in its own right, with its own concepts, methods, techniques, and language. The *Encyclopedia Brittanica*, in dividing up knowledge, lists logic as one of the seven branches of knowledge. More recently, the study of logic has played a major role in the development of modern day computers and programming languages. Logic continues to play an important part in computer science; indeed, it has been said that computer science is logic plus electrical engineering.

This book is intended to introduce you to some of the most important concepts and tools of logic. Our main aim is to provide detailed and systematic answers to the questions raised above. We want you to understand just how the laws of logic follow inevitably from the meanings we associate with the language used to make claims. Convention is crucial in establishing the meaning of a language, but once the meaning is established, the laws of logic inevitably follow.

More particularly, we have two main aims. The first is to help you learn a new language, the language of first-order logic. Our second aim is to help you learn about the notion of logical consequence, and about how one goes about establishing whether some claim is or is not a logical consequence of other accepted premises. While there is much more to logic than we can even hint at in this book, or than any one person could learn in a lifetime, we can at least cover these most basic of issues.

1.2 Why learn an artificial language?

The language FOL mentioned just now is very important. Like Latin, this language is not spoken, but unlike Latin, it is used every day by mathematicians, philosophers, computer scientists, linguists, and practitioners of artificial intelligence. Indeed, in some ways it is the *lingua franca* of the symbolic sciences.

The language goes by various names: the lower predicate calculus, the functional calculus, the language of first-order logic, and FOL. The last of these is pronounced *ef–oh–el*, not *fall*, and is the name we will use.

Certain elements of FOL go back to Aristotle, but the language as we know it today has emerged over the past hundred years. The names chiefly associated with its development are those of Gottlob Frege, Giuseppe Peano, and Charles Sanders Peirce. In the late 19th century, these three logicians independently came up with the most important elements of the language, known as the quantifiers. Since then, there has been a process of standardization and simplification, resulting in the language in its present form. Even so, there remain certain dialects of FOL, differing mainly in the choice of symbols used for the connectives and quantifiers. We will use the dialect most common in mathematics, though we will also introduce several other dialects along the way.

FOL is used in different ways in different fields. In mathematics, FOL is used in an informal way quite extensively. The various connectives and quantifiers find their way into a great deal of mathematical discourse, both formal and informal, as in a classroom setting. Here you will often find elements of FOL interspersed with English or the mathematician's native language. What student of calculus has not seen such formulas as:

$$\forall\epsilon > 0 \ \exists\delta > 0 \ldots$$

Here, the unusual, rotated letters are taken directly from the language FOL.

In philosophy, FOL and enrichments of it are used in two different ways. As in mathematics, the notation of FOL is used when absolute clarity, rigor, and lack of ambiguity are essential. But it is also used as a case study of informal notions (like grammaticality, meaning, truth, and proof) being made precise and rigorous. The uses in linguistics stem from this use, since linguistics is concerned, in large part, with understanding some of these same informal notions.

In artificial intelligence, FOL is also used in two ways. Some researchers take advantage of the simple structure of FOL sentences to use it as a way to encode knowledge to be stored and used by a computer. Thinking is modeled by manipulations involving sentences of FOL. Another use is as a precise specification language for stating axioms and proving results about would-be robots.

In computer science, FOL has had an even more profound effect. The very idea of an artificial language that is precise yet rich enough to program computers was inspired by FOL. In addition, all extant programming languages borrow some notions from one or another dialect of FOL. Finally, there are so-called logic programming languages, like Prolog, whose programs are sequences of sentences in a certain dialect of FOL. We will discuss the logical basis of Prolog a bit in Part IV of this book.

FOL serves as the prototypical example of what is known as an artificial language. These are languages that were designed for special purposes, and are contrasted with so-called natural languages, languages like English and Greek that people actually speak. The design of artificial languages within the symbolic sciences is an important activity, one that is based on the success of FOL and its descendants.

Even if you are not going to pursue logic or any of the symbolic sciences, the study of FOL can be of real benefit. That is why it is so widely taught. For one thing, learning FOL is an easy way to demystify a lot of formal work. But more importantly, it will teach you a great deal about your own language, and the laws of logic it supports. First, FOL, while very simple, incorporates in a clean way some of the important features of human languages. This helps make these features much more transparent. Chief among these is the relationship between language and the world. But, secondly, as we attempt to translate English sentences into FOL we will also gain an appreciation of the great subtlety that resides in English, subtlety that cannot be captured in FOL or similar languages, at least not yet. Finally, we will gain an awareness of the enormous ambiguity present in almost every English sentence, ambiguity which somehow does not prevent us from understanding each other in most situations.

Earlier, we asked what makes one claim follow from others: convention, or something else? Giving an answer to this question for FOL takes up a significant part of this book. But a short answer can be given here. Modern logic teaches us that one claim is a logical consequence of another if there is no way the latter could be true without the former also being true.

This is the notion of logical consequence implicit in all rational inquiry. All the rational disciplines implicitly presuppose that this notion makes sense, and that we can use it to extract consequences of what we know to be so, or what we think might be so. It is also used in disconfirming a theory. For if a claim S is a logical consequence of a theory, and we discover that S is false, then we know the theory itself must be false. One of our major concerns in this book is to examine this notion of logical consequence as it applies specifically to the language FOL. We will study methods of proof—how we can *prove* that one sentence of FOL is a logical consequence of another—and also methods for showing that one claim is *not* a logical consequence of others.

1.3 About this book

This book, and the supporting computer program that comes with it, grew out of our own dissatisfaction with beginning logic courses. It seems to us that students are all too often coming away from these courses with neither of the things we want them to have. Indeed, they often come away from such classes with exactly the opposite from the truth about logic. They leave a first (and only) course in logic having learned what seem like a bunch of useless formal rules. They have little if any understanding about why those rules, rather than some others, are used. And they are unable to take any of what they have learned and use it in other fields of rational inquiry. Indeed, many come away convinced that logic is both arbitrary and irrelevant.

The real problem, as we see it, is a failure on the part of us logicians to find a simple way to explain the relationship between meaning and the laws of logic. In particular, we do not succeed in conveying to students what sentences in FOL mean, or in conveying how the meanings of sentences govern which methods of inference are valid and which are not. It is this problem we set out to solve with this book and computer program.

There are two ways to learn a second language. One is to learn how to translate sentences of the language to and from sentences of your native language. The other way is to learn by using the language directly. In teaching FOL, the first way has always been the prevailing method of instruction. There are serious problems with this approach. Some of the problems, oddly enough, stem from the simplicity, precision, and elegance of FOL. This results in a distracting mismatch between the student's native language and FOL. It forces students trying to learn FOL to be sensitive to subtleties of their native language that normally go unnoticed. While this is useful, it often interferes with the learning of FOL. Students mistake complexities of their native tongue for complexities of the new language they are learning.

In this book, we adopt the second method for learning FOL. You will be given many tasks involving the language, tasks that will help you understand the meanings of sentences in FOL. Only then will you be asked to translate between English and FOL. Correct translation involves finding a sentence in the target language whose meaning approximates, as closely as possible, the meaning of the sentence being translated. To do this well, a translator must already be fluent in both languages.

We have been using this approach for several years. What allows it to work is Tarski's World, the computer program that accompanies this book. Tarski's World provides a simple environment in which FOL can be used in many of the ways that we use our native language. We provide a large number of problems and exercises that walk you through the use of the language in this setting. We build on this in other problems where you will learn how to put the language to more sophisticated uses.

As we said earlier, besides teaching the language FOL, we also discuss basic methods of proof and how to use them. In this regard, too, our approach is somewhat unusual. We concentrate on informal methods of proof. Though we present a Fitch-style deductive system $\mathcal{F}$ as a mathematical formalization of these informal methods, the book can be used without covering the sections in which $\mathcal{F}$ is presented. Students will still learn the basic techniques for giving rigorous proofs.

Tarski's World also plays a major role in our discussion of proof, by providing an environment for showing that one claim does not follow from another. One learns not just how to prove consequences of premises, but the equally important technique of showing that a given claim does not follow logically from the premises. One simply finds a set of circumstances (a "world") in which the premises are true but the purported conclusion false.

Besides the two basic aims mentioned earlier, we also have a number of other points to make in this book. One results in our emphasis on languages where all the basic symbols are assumed to be meaningful. This is in contrast to the so-called "uninterpreted languages" (surely an oxymoron) so often found in logic textbooks. Another is that, as logic becomes more important in computer science, there are a number of important topics that deserve to be presented the first time around. This belief has resulted in the inclusion of various such topics in this book. They are contained in Part IV, which can be skipped at the discretion of the instructor.

Part I

Propositional Logic

2

Atomic Sentences

In Chapter 1, we talked about FOL as though it were a single language. Actually, it is more like a family of languages, all having a similar grammar and sharing certain important vocabulary items, known as the connectives and quantifiers. Languages in this family can differ, however, in the specific vocabulary used to form their most basic sentences, the so-called atomic sentences.

Atomic sentences correspond to the most simple sentences of English, sentences consisting of some names connected by a predicate. Examples are *Max ran, Max saw Claire,* and *Claire gave Scruffy to Max.* In FOL, atomic sentences are also formed by combining names (or individual constants, as they are often called) and predicates, though the way they are combined is a bit different, as you will see.

Different versions of FOL have available different names and predicates. The first-order language used in Tarski's World takes the letters a through f plus n1, n2, ... as its names. (You are in charge of the first six, in that you can freely use them to name objects. Tarski's World is in charge of the remaining names, which it uses when playing the game with you.) For predicates, Tarski's World uses Tet, Cube, Dodec, Small, Medium, Large, =, Smaller, Larger, LeftOf, RightOf, BackOf, FrontOf, and Between. Some examples of atomic sentences in this language are Cube(b), Larger(c, f), and Between(b, c, d). These sentences say, respectively, that b is a cube, that c is larger than f, and that b is between c and d.

Later in this chapter, we will look at the atomic sentences used in two other versions of FOL, the first-order languages of set theory and arithmetic. In the next chapter, we begin our discussion of the connectives and quantifiers common to all first-order languages.

2.1 Individual constants

Individual constants are simply symbols that are used to refer to some fixed individual object or other. They are the FOL analogue of names.

For example, we might use Max as an individual constant to denote a particular person, or 1 as an individual constant to denote a particular number. In either case, they would basically work exactly the way names work in English.

The main difference between names in English and the individual constants of FOL is that we require the latter to refer to exactly one object. Obviously, the name *Max* in English can be used to refer to many different people, and might even be used twice in a single sentence to refer to two different people. Such wayward behavior is frowned upon in FOL.

There are also names in English that do not refer to any actually existing object. For example *Pegasus*, *Zeus*, and *Santa Claus* are perfectly fine names in English; they just fail to refer to anything or anybody. We don't allow such names in FOL.[1] What we do allow, though, is for one object to have more than one name; thus the names Matthew and Max might both refer to the same individual. We also allow for nameless objects, objects that have no name at all.

Remember

In FOL,

- Every name must name an object.
- No name can name more than one object.
- An object can have more than one name, or no name at all.

2.2 Predicate symbols

Predicate symbols are symbols used to denote some property of objects or some relation between objects. As in English, they are expressions that, when combined with names, form atomic sentences. But they don't correspond exactly to the predicates of English grammar.

Consider the English sentence *Max likes Claire.* In English grammar, this is analyzed as a subject-predicate sentence. It consists of the subject *Max* followed by the predicate *likes Claire.* In FOL, by contrast, we view this as a claim involving two "logical subjects," the names *Max* and *Claire*, and a predicate, *likes,* that expresses a relation between the referents of the names. Thus atomic sentences of FOL often have two or more logical subjects, and the predicate is, so to speak, whatever is left. The logical subjects are called the "arguments" of the predicate. In this case, the predicate is said to be binary, since it takes two arguments.

[1]There is, however, a variant of first-order logic called *free logic* in which this assumption is relaxed. In free logic, there can be individual constants without referents. This yields a language more appropriate for mythology and fiction.

In English, some predicates have optional arguments. Thus you can say *Claire gave, Claire gave Scruffy,* or *Claire gave Scruffy to Max.* Here the predicate *gave* is taking one, two, and three arguments, respectively. But in FOL, each predicate has a fixed number of arguments, a fixed *arity* as it is called. This is a number that tells you how many individual constants the predicate symbol needs in order to form a sentence. The term "arity" comes from the fact that predicates taking one argument are called *unary*, those taking two are *binary*, and so forth.

If the arity of a predicate symbol Pred is 1, then Pred will be used to denote some property of objects, and so will require exactly one argument (a name) to make a claim. For example, we might use the unary predicate symbol AtHome to denote the property of being at home. We could then combine this with the name Max to get the expression AtHome(Max), which expresses the claim that Max is at home.

If the arity of Pred is 2, then Pred will be used to represent a relation between two objects. Thus, we might use the expression Taller(Claire, Max) to express a claim about Max and Claire, the claim that Claire is taller than Max. In FOL, we can have predicate symbols of any arity. However, in the language of Tarski's World we restrict ourselves to predicates with arities 1, 2, and 3. Here we list again the predicates of that language, this time with their arity.

Arity 1: Cube, Tet, Dodec, Small, Medium, Large
Arity 2: Smaller, Larger, LeftOf, RightOf, BackOf, FrontOf, =
Arity 3: Between

Tarski's World assigns each of these predicates a fixed interpretation, one reasonably consistent with the corresponding English verb phrase. For example, Cube corresponds to *is a cube*, BackOf corresponds to *is in back of*, and so forth. You can get the hang of them by working through the first set of exercises given below.

In English, predicates are sometimes vague. It is often unclear whether an individual has the property in question or not. For example Claire, who is six, is young. She will not be young when she is 96. But there is no determinate age at which a person stops being young: it is a gradual sort of thing. FOL, however, assumes that every predicate is interpreted by a determinate property or relation. By a *determinate* property, we mean a property for which, given any object, there is a definite fact of the matter whether or not the object has the property.

This is one of the reasons we say that the Tarski's World predicates are *somewhat* consistent with the corresponding English predicates. Unlike the English predicates, they are given very precise interpretations, interpretations that are suggested by, but not necessarily identical with, the meanings of the corresponding English phrases. The case where the discrepancy is probably the greatest is between Between and *is between.*

Remember

In FOL,

- Every predicate symbol comes with a single, fixed "arity," a number that tells you how many names it needs to form an atomic sentence.
- Every predicate is interpreted by a determinate property or relation of the same arity as the predicate.

2.3 Atomic sentences

In FOL, the simplest kinds of claims are those made with a single predicate and the appropriate number of individual constants. A sentence formed by a predicate followed by the right number of names is called an *atomic* sentence. For example Taller(Claire, Max) and Cube(a) are atomic sentences, provided the names and predicate symbols in question are part of the vocabulary of our language. In the case of the identity symbol, we put the two required names on either side of the predicate, as in $a = b$. This is called "infix" notation, since the predicate symbol $=$ appears in between its two arguments. With the other predicates we use "prefix" notation: the predicate precedes the arguments.

The order of the names in an atomic sentence is quite important. Just as *Claire is taller than Max* means something different than *Max is taller than Claire,* so too Taller(Claire, Max) means something different than Taller(Max, Claire). We have set things up in Tarski's World so that the order of the arguments of our predicates is like that in English. Thus LeftOf(b, c) means more or less the same thing as the English sentence *b is left of c,* and Between(b, c, d) means roughly the same as the English *b is between c and d.*

Predicates and names refer to properties and objects, respectively. What makes sentences special is that they make claims (or express propositions). A claim is something that is either true or false; which of these it is we call its *truth value.* Thus Taller(Claire, Max) expresses a claim whose truth value is TRUE, while Taller(Max, Claire) expresses a claim whose truth value is FALSE. (You probably didn't know that, but now you do.) Given our assumption that predicates express determinate properties and that names denote definite individuals, it follows that each atomic sentence of FOL must express a claim that is either true or false.

Remember

In FOL,

- Atomic sentences are formed by putting a predicate of arity n in front of n names (enclosed in parentheses and separated by commas).
- Atomic sentences are built from the identity predicate, =, using infix notation: the arguments are placed on either side of the predicate.
- The order of the names is crucial in forming atomic sentences.

Exercises and Problems

You will eventually want to read all of Appendix A, on how to use Tarski's World. To do the following problems, you will need to read the first four sections, through section A4.

Exercise$_\Diamond$ 1. (Understanding atomic sentences) Open the files called Wittgenstein's World and Wittgenstein's Sentences. Notice that all the sentences in this file are atomic sentences. Run through them assessing their truth values in the given world. Use the evaluation capability of Tarski's World to check your answers. (Since the sentences are all atomic sentences the game will not be helpful.) Then change Wittgenstein's World in many different ways, seeing what happens to the truth of the various sentences. The main point of this exercise is to help you figure out how Tarski's World interprets the various predicates. For example, what does BackOf(d, c) mean? Do two things have to be in the same column for one to be in back of the other? Play around as much as you need until you are sure you understand the meanings of the atomic sentences in this file. When you are finished, do not save the changes you have made to the files.

In doing this exercise, you will no doubt notice that Between does not mean exactly what the English *between* means. This is due to the necessity of interpreting Between as a determinate predicate. We have chosen the following rule: To find out whether b is between c and d, draw a line connecting the squares that contain c and d. If c and d lie either on the same row or column or on adjacent rows or columns, then draw the line from the midpoints of the containing squares' facing sides; otherwise, draw it from the closest corners of the squares. If the square containing b intersects this line at some point other than an endpoint, then b is considered between c and d.

Problem$_\Diamond$ 2. (Copying some atomic sentences) This exercise will give you some practice with the Tarski's World Keyboard, as well as with the syntax of atomic sentences. The following are all atomic sentences of our language. Start a new sentence file and copy them into it. Using the Evaluation box, check each formula after you write it to see that it is a sentence. If you make a mistake, edit it before going on.

1. Tet(a)
2. Medium(a)
3. Dodec(b)
4. Cube(c)
5. FrontOf(a, b)
6. Between(a, b, c)
7. a = d
8. Larger(a, b)
9. Smaller(a, c)
10. LeftOf(b, c)

Save these sentences in a file named Sentences 2.2.[2]

Problem$_\diamond$ 3. (Building a world) Build a world in which all the sentences in Problem 2 are simultaneously true. Save it in a file named World 2.3.

Problem$_\diamond$ 4. (Translating atomic sentences) Here are some simple sentences of English. Start a new sentence file and translate them into FOL.

1. *a is a cube.*
2. *b is smaller than a.*
3. *c is between a and d.*
4. *d is large.*
5. *e is larger than a.*
6. *b is a tetrahedron.*
7. *e is a dodecahedron.*
8. *e is right of b.*
9. *a is smaller than e.*
10. *d is in back of a.*

Now build a world in which all of them are true. Name these files Sentences 2.4 and World 2.4.

[2]When we ask you to save a file using a specific name, it is important that you use exactly the name suggested, for a couple of reasons. First of all, we will sometimes ask you to go back and open these files, and if they are not named what they should be, you may have problems finding them. Second, if your instructor is collecting homework problems on disk, then the Tarski's World grading facility will need the right name to find the file. For example, it will not be able to find the file for this exercise if you name it Sentences (2.2) or Sentences–2.2 rather than Sentences 2.2.

Problem$_\diamond$ 5. (Naming objects) Open Lestrade's Sentences and Lestrade's World. You will notice that none of the objects in this world has a name. Your task is to assign them names in such a way that all the sentences in the list come out true. Save the resulting world in a file named World 2.5. Be sure to use **Save World As...**, not **Save World**, as explained in Appendix A.

Problem* 6. (Naming objects, continued) How many of the choices in Problem 5 were forced on you? That is, how many of the names could you have assigned to other objects and still had all the sentences come out true?

Problem$_\diamond$ 7. (Context of predicates) We have stressed the fact that FOL assumes that every predicate is interpreted by a determinate relation, whereas this is not the case in natural languages like English. Indeed, even when things seem quite determinate, there is often some form of context sensitivity. In fact, we have built some of this into Tarski's World. Consider, for example, the difference between the predicates Larger and BackOf. Whether or not cube a is larger than cube b is a determinate matter, and also one that does not vary depending on your perspective on the world. Whether or not a is back of b is also determinate, but, in this case, it does depend on your perspective. If you rotate the world by 90°, the answer might change.

Open Austin's Sentences and Wittgenstein's World. Evaluate each sentence and tabulate the resulting truth values on a piece of paper. Then rotate the world 90° to the right and evaluate the sentences again, adding the results to your table. Repeat this until the world has come full circle.

Do you see any patterns in your table? In particular, suppose you were given the following row from such a table:

FALSE TRUE FALSE FALSE

What could you say about the predicate in the corresponding atomic sentence? What could you say about the relative positions of the objects named in the sentence?

2.4 The first-order language of set theory

FOL was initially developed for use in mathematics, and consequently the most familiar first-order languages are those associated with various branches of mathematics. One of the most common of these is the language of set theory. This language has only two predicates, both binary. The first is the identity symbol, =, which we have already encountered, and the second is the symbol $\in$, for set membership.

It is standard to use infix notation for both of these predicates. Thus in set theory, atomic sentences are always formed by placing individual constants on either side of one of the two predicates. This allows us to make identity claims, of the form a = b, and membership claims, of the form a ∈ b (where a and b are individual constants).

A sentence of the form a ∈ b is true if and only if the thing named by b is a set, and the thing named by a is a member of that set. For example, suppose a names the number 2 and b names the set $\{2, 4, 6\}$. Then the following table tells us which membership claims made up using these names are true and which are false.[3]

a ∈ a	FALSE
a ∈ b	TRUE
b ∈ a	FALSE
b ∈ b	FALSE

Notice that there is one striking difference between the atomic sentences of set theory and the atomic sentences of Tarski's World. In Tarski's World, you can have a sentence, like LeftOf(a, b), that is true in a world, but which can be made false simply by moving one of the objects. Moving an object does not change the way the name works, but it can turn a true sentence into a false one, just as the sentence *Claire is sitting down* can go from true to false in virtue of Claire's standing up.

In set theory, we won't find this sort of thing happening. Here, the analog of a world is just a domain of objects and sets. For example, our domain might consist of all natural numbers, sets of natural numbers, sets of sets of natural numbers, and so forth. The difference between these "worlds" and those of Tarski's World is that the truth or falsity of the atomic sentences is determined entirely once the reference of the names is fixed. There is nothing that corresponds to moving the blocks around. Thus if the universe contains the objects 2 and $\{2, 4, 6\}$, and if the names a and b are assigned to them, then the atomic sentences must get the values indicated in the previous table. The only way those values can change is if the names name different things. Identity claims also work this way, both in set theory and in Tarski's World.

Exercises and Problems

Problem 8. Which of the following atomic sentences in the first-order language of set theory are true and which are false? We use, in addition to a and b as above, the name c for 6 and d for $\{2, 7, \{2, 4, 6\}\}$.

1. a ∈ c
2. a ∈ d
3. b ∈ c

[3]For the purposes of this discussion we are assuming that numbers are not sets, and that sets can contain either numbers or other sets as members.

4. b ∈ d
5. c ∈ d
6. c ∈ b

2.5 Function symbols

Some first-order languages have, in addition to names and predicates, other expressions that can appear in atomic sentences. These expressions are called *function symbols.* Function symbols allow us to form name-like terms from names and other name-like terms. Some English examples will help clarify this.

English has many sorts of noun phrases, expressions that can be combined with a verb phrase to get a sentence. Besides names like *Max* and *Claire*, other noun phrases include expressions like *Max's father*, *Claire's mother*, *Every girl who knows Max*, *No boy who knows Claire*, *Someone* and so forth. Each of these combines with a singular verb phrase such as *likes unbuttered popcorn* to make a sentence. But notice that the sentences that result have very different logical properties. For example,

Claire's mother likes unbuttered popcorn

implies that someone likes unbuttered popcorn, while

No boy who knows Claire likes unbuttered popcorn

does not.

Since these noun phrases have such different logical properties, they are treated differently in FOL. Those that intuitively refer to an individual are called "terms," and behave like the individual constants we have already discussed. In fact, individual constants are the simplest terms, and more complex terms are built from them using function symbols. Noun phrases like *No boy who knows Claire* are handled with very different devices, known as quantifiers, which we will discuss later.

The FOL analog of the noun phrase *Max's father* is the term father(Max). It is formed by putting a function symbol, father, in front of the name Max. The result is a complex term that we use to refer to the father of the person referred to by the name Max. Similarly, we can put the function symbol mother together with the name Claire and get the term mother(Claire), which functions pretty much like the English term *Claire's mother.*

We can repeat this construction as many times as we like, forming more and more complex terms:

father(father(Max)),
mother(father(Claire)),
mother(mother(mother(Claire))),

The first of these refers to Max's paternal grandfather, the second to Claire's paternal grandmother, and so forth.

These function symbols are called unary function symbols, because, like unary predicates, they take one argument. The resulting terms function just like names, and can be used in forming atomic sentences. For instance, the FOL sentence

TallerThan(father(Max), Max)

says that Max's father is taller than Max.

Students often confuse function symbols with predicates, because both take terms as arguments. But there is a big difference. When you combine a unary function symbol with a term you do not get a sentence, but another term: something that refers (or should refer) to an object of some sort. This is why function symbols can be reapplied over and over again. As we have seen, the following makes perfectly good sense:

father(father(Max)).

This, on the other hand, is total nonsense:

Dodec(Dodec(a)).

Besides unary function symbols, FOL allows function symbols of any arity. Thus, for example, we can have binary function symbols. Simple English counterparts of binary function symbols are hard to come up with, but they are quite common in mathematics. For instance, we might have a function symbol sum that combines with two terms, t_1 and t_2, to give a new term, $sum(t_1, t_2)$, which refers to the sum of the numbers referred to by t_1 and t_2. Then the complex term $sum(3, 5)$ would give us another way of referring to 8. In the next section, we will introduce a function symbol to denote addition, but we will use infix notation, rather than prefix notation. Thus $3 + 5$ will be used instead of $sum(3, 5)$.

Remember

In a language with function symbols,

- Complex terms are typically formed by putting a function symbol of arity n in front of n terms (simple or complex).
- Complex terms are used just like names (simple terms) in forming atomic sentences.

Exercises and Problems

Problem 9. Suppose we have two first-order languages for talking about fathers. The first, which we'll call the functional language, contains the names Claire, Melanie, and Jon, the function symbol father, and the predicates = and TallerThan. The second language, which we will call the relational language, has the same names, no function symbols, and the binary predicates TallerThan and FatherOf, where FatherOf(c, b) means that c is the father of b. Translate the following atomic sentences from the relational language into the functional language. Be careful. Some atomic sentences, such as Claire = Claire, are in both languages! Such a sentence counts as a translation of itself.

1. FatherOf(Jon, Claire)
2. FatherOf(Jon, Melanie)
3. TallerThan(Melanie, Claire)

Which of the following atomic sentences of the functional language can be translated into atomic sentences of the relational language?

1. father(Melanie) = Jon
2. father(Melanie) = father(Claire)
3. TallerThan(father(Claire), father(Jon))

When we add connectives and quantifiers to the language, we will be able to translate freely back and forth between the functional and relational languages.

Problem 10. Make up a first-order language for talking about people and their relative heights. Instead of using relation symbols like TallerThan, however, use a function symbol that allows you to refer to people's heights, plus the relation symbols = and <. Your language should allow you to express such claims as *Carl is taller than Sam* and *Sam and Mary are the same height.*

2.6 The first-order language of arithmetic

Neither the language of Tarski's World nor the language of set theory has function symbols. One first-order language that uses them extensively is the language of arithmetic. This language allows us to express statements about the natural numbers $0, 1, 2, 3, \ldots$, and the arithmetic operations of addition and multiplication.

There are several more or less equivalent ways of setting up this language. The one we will use has two names, 0 and 1, two binary relation symbols, = and <, and two binary function symbols, + and ×. The atomic sentences are those that can be built up out of these symbols. We will use infix notation both for the relation symbols and the function symbols.

Notice that there are an infinite number of different terms in this language (for example, 0, 1, $(1+1)$, $((1+1)+1)$, $(((1+1)+1)+1)$, ...), and so an infinite number of atomic sentences. Our example also shows that every natural number is named by some term of the language. This raises the question of how we can specify the set of terms in a precise way. We can't list them, since there are too many. The way we get around this is by using what is known as an *inductive* definition.

Definition 1. The terms of first-order arithmetic are formed in the following way:

1. The names 0, 1 are terms.
2. If t_1, t_2 are terms, then the expressions $(t_1 + t_2)$ and $(t_1 \times t_2)$ are also terms.
3. Nothing is a term unless it can be obtained by repeated application of (1) and (2).

We should point out that this definition does indeed allow the function symbols to be applied over and over. Thus $(1 + 1)$ is a term by clause 2 and the fact that 1 is a term. In which case $((1 + 1) \times (1 + 1))$ is also a term, again by clause 2. And so forth.

In Chapter 9, we will see how to give a definition like the above in a more satisfactory way, one that avoids the somewhat vague third clause.

The atomic sentences in the language of first-order arithmetic are those that can be formed from the terms and the two binary predicate symbols, $=$ and $<$. So, for example, the FOL version of *1 times 1 is less than 1 plus 1* is the following:

$$(1 \times 1) < (1 + 1)$$

Exercises and Problems

Problem 11. Show that the following expressions are terms in the first-order language of arithmetic. What numbers do they refer to?

1. $(0 + 0)$
2. $(0 + (1 \times 0))$
3. $((1 + 1) + ((1 + 1) \times (1 + 1)))$
4. $(((1 \times 1) \times 1) \times 1)$

Problem 12. Find a way to express the fact that three is less than four.

Problem* 13. Show that there are infinitely many terms that refer to the number one.

2.7 General first-order languages

In general, a first-order language is specified by fixing the names, predicates, and function symbols it contains. Each predicate and function symbol comes with a specified arity.

A first-order language must contain at least one predicate symbol, although sometimes = is the only predicate symbol used. But it need not contain any function symbols. As we have seen, neither the language of Tarski's World nor the language of set theory uses function symbols.

It turns out, surprisingly enough, that some first-order languages do not contain any names, either. Since we haven't explained how you can form sentences without names, this may seem a bit puzzling. But as a matter of fact, the language of set theory often gets by with no individual constants: a point we deftly sneaked past in our earlier discussion. We will see how this is possible when we get to quantifiers.

First-order languages differ in the names, predicates, and function symbols they contain, and so in the atomic sentences that can be formed. What they share in common are the connectives and quantifiers that enable us to build more complex sentences from these simpler parts. We will get to those common elements in later chapters.

When you translate a sentence of English into FOL, you will sometimes have a "predefined" first-order language that you want to use, like the language of Tarski's World or the language of arithmetic described in the last section. If so, your goal is to come up with a translation that captures the meaning of the original English sentence as nearly as possible, given the names, predicates, and function symbols available in your first-order language.

Other times, though, you will not have a predefined language to use for your translation. If not, the first thing you have to do is decide what names, predicates, and functions you need for your translation. In effect, you are designing, on the fly, a new first-order language capable of expressing the English sentence you want to translate. We've been doing this all along, for example when we introduced AtHome(Max) as the translation of *Max is at home* and Taller(Claire, Max) as the translation of *Claire is taller than Max.*

When you make these decisions, there are often alternative ways to go. For example, suppose you were asked to translate the sentence *Claire gave Scruffy to Max.* You might introduce a binary predicate GaveScruffy(x, y), meaning *x gave Scruffy to y*, and then translate the original sentence as GaveScruffy(Claire, Max). Alternatively, you might introduce a three-place predicate Gave(x, y, z), meaning *x gave y to z*, and then translate the sentence as Gave(Claire, Scruffy, Max).

There is nothing wrong with either of these predicates, or their resulting translations, so long as you have clearly specified what the predicates mean. Of course, they may not be equally useful when you go on to translate other sentences. The first predicate will allow you to translate sentences like *Max gave Scruffy to Evan* and *Evan gave Scruffy to Miles.* But if you then run into the sentence *Max gave Carl to Claire,* you would be stuck, and would have to introduce an entirely new predicate, say, GaveCarl(x, y). The three-place predicate is thus more flexible. A first-order language that contained it (plus the relevant names) would be able to translate any of these sentences.

In general, when designing a first-order language we try to economize on the predicates by introducing more flexible ones, like Gave(x, y, z), rather than less flexible ones, like GaveScruffy(x, y) and GaveCarl(x, y). This produces a more expressive language, and one that makes the logical relations between various claims more perspicuous.

Names can be introduced into a first-order language to refer to anything that can be considered an object. But we construe the notion of an "object" pretty flexibly—to cover anything that we can make claims about. We've already seen languages with names for people, sets, numbers, and the blocks of Tarski's World. Sometimes we will want to have names for still other kinds of "objects," like days or times. Suppose, for example, that we want to translate the sentences:

Claire gave Scruffy to Max on Saturday.
Sunday, Max gave Scruffy to Evan.

Here, we might introduce a four-place predicate Gave(w, x, y, z), meaning *w gave x to y on day z*, plus names for particular days, like last Saturday and last Sunday. The resulting translations would look something like this:

Gave(Claire, Scruffy, Max, Saturday)
Gave(Max, Scruffy, Evan, Sunday)

Designing a first-order language with just the right names, predicates, and function symbols requires some skill. Usually, the overall goal is to come up with a language that can say everything you want, but that uses the smallest "vocabulary" possible. Picking the right names, predicates, and function symbols is the key to doing this.

Exercises and Problems

Problem 14. Suppose we have two first-order languages: the first contains the binary predicates GaveScruffy(x, y) and GaveCarl(x, y), and the names Max and Claire; the second contains the ternary predicate Gave(x, y, z) and the names Max, Claire, Scruffy, and Carl.

1. List all of the atomic sentences that can be expressed in the first language. (Some of these may say slightly odd things, like GaveScruffy(Claire, Claire), but don't worry about that.)
2. How many atomic sentences can be expressed in the second language? (Count all of them, including odd ones like Gave(Scruffy, Scruffy, Scruffy).)

TABLE 1
Names and predicates for a language

ENGLISH	FOL	COMMENT
Names:		
Max	Max	
Claire	Claire	
Folly	Folly	The name of a certain disk.
Silly	Silly	The name of another disk.
2 pm, Jan 2, 1993	2:00	The name of a time.
2:01 pm, Jan 2, 1993	2:01	One minute later.
⋮	⋮	Similarly for other times.
Predicates:		
x is a disk	Disk(x)	
x is a person	Person(x)	
x is a student	Student(x)	
t is earlier than t′	t < t′	Earlier-than for times.
x was blank at time t	Blank(x, t)	
x was angry at time t	Angry(x, t)	
x owned y at time t	Owned(x, y, t)	
x gave y to z at t	Gave(x, y, z, t)	
x erased y at time t	Erased(x, y, t)	

3. How many names and binary predicates would a language like the first need in order to say everything you can say in the second?

Problem 15. We will be giving a number of problems that use the symbols explained in Table 1. Using this table, translate the following sentences into FOL. All references to times are assumed to be to times on January 2, 1993.

1. *Claire owned Folly at 2 pm.*
2. *Claire gave Silly to Max at 2:05 pm.*
3. *Max is a student.*
4. *Claire erased Folly at 2 pm.*
5. *Folly belonged to Max at 3:05 pm.*
6. *2:00 pm is earlier than 2:05 pm.*

Problem 16. Translate the following into colloquial English, consulting Table 1.

1. Owned(Max, Silly, 2:00)
2. Erased(Max, Silly, 2:30)

3. Gave(Max, Silly, Claire, 2:00)
4. (2:00 < 2:00)

Problem* 17. For each sentence in the following list, suggest a translation into an atomic sentence of FOL. In addition to giving the translation, explain what kinds of objects your names refer to and the intended meaning of the predicate you use. If you use any function symbols, explain their meanings as well.

1. *AIDS is less contagious than influenza.*
2. *Spain is between France and Portugal in size.*
3. *Misery loves company.*
4. *Max shook Claire's father's hand.*
5. *John and Nancy's eldest child is younger than Jon and Mary Ellen's.*

2.8 Methods of proof

A major concern in logic is the concept of *logical consequence*: When does one sentence, statement, or claim follow logically from some others? In fact, one of the main motivations in the design of FOL was to make the logical consequence relation as perspicuous as possible. By avoiding the ambiguities and complexities of ordinary language, it was hoped that the consequences of our claims would be more easily recognizable.

Just what do we mean by logical consequence? Or rather, since this phrase is sometimes used in quite different contexts, what does a *logician* mean by logical consequence? A few examples will help. First, let's say that an *argument* is any series of statements in which one (called the *conclusion*) is meant to follow from, or be supported by, the others (called the *premises*). Here are a couple of examples:

> All men are mortal. Socrates is a man. So, Socrates is mortal.
>
> Lucretius is a man. After all, Lucretius is mortal and all men are mortal.

One difference between these two arguments is the placement of the conclusion. In the first argument, the conclusion comes at the end, while in the second, it comes at the start. A more important difference is that the first argument is good, while the second is bad. We will say that the first argument is *logically valid*, or that its conclusion is a *logical consequence* of its premises. The reason we say this is that it is impossible for this conclusion to be false if the premises are true. In contrast, our second conclusion might be false (suppose Lucretius is my pet goldfish), even though the premises are true (goldfish are notoriously mortal).

Roughly speaking, an argument is logically valid if and only if the conclusion must be true on the assumption that the premises are true. Notice that this does not mean that an argument's premises have to be true in order for it to be valid. Our first example above would be a valid argument even if it turned out that we were mistaken about one of the premises, say if Socrates turned out to be a robot rather than a man. It would still be impossible for the premises to be true and the conclusion false. In that eventuality, we would say that the argument was valid, but since it had a false premise, we would still not be guaranteed that the conclusion was true.

Here is another example of a valid argument, this time one expressed in the language of Tarski's World. Suppose we are told that Cube(c) and that c = b. Then it certainly follows that Cube(b). Why? Because there is no possible way for the premises to be true—for c to be a cube and for c to be the very same object as b—without the conclusion being true as well. Note that we can recognize that the last statement is a consequence of the first two without knowing that the premises are actually, as a matter of fact, true. For the crucial observation is that *if* the premises are true, *then* the conclusion must also be true.

This description of the logical consequence relation is fine, as far as it goes. But it doesn't give us everything we would like. In particular, it does not tell us how to *show* that a given conclusion S follows, or does not follow, from some premises P, Q, R, ... In the examples we have looked at, this may not seem very problematic, since the answers are obvious. But when we are dealing with more complicated sentences or more subtle reasoning, things are not so simple.

In this book, we will discuss some basic methods of showing when claims follow from other claims and when they do not. The main technique for doing the latter, for showing that a given conclusion does *not* follow from some premises, is to find a possible circumstance in which the premises are true but the conclusion false. In fact we have already used this method once, to show that our argument about Lucretius is invalid. We will use this method repeatedly, and introduce more precise versions of it as we go on.

What methods are available to us for showing that a given claim *is* a logical consequence of some premises? Here, the key notion is that of a *proof*. A proof is a step-by-step demonstration that a given conclusion (say S) follows from some premises (say P, Q, R). The way a proof works is by establishing a series of intermediate conclusions, each of which is an obvious consequence of the original premises and the intermediate conclusions previously established. The proof ends when we finally establish S as an obvious consequence of the original premises and the intermediate conclusions. For example, from P, Q, R it might be obvious that S_1 follows. And from all of these, including S_1, it might be obvious that S_2 follows. Finally, from all these together we might be able to draw our desired conclusion S. If our individual steps are correct, then the proof shows that S is indeed a consequence of P, Q, R. After all, if the premises are all true, then our intermediate conclusions must be true as well. And in that case, our final conclusion must be true, too.

Consider a simple, concrete example. Suppose we want to show that *Socrates sometimes worries about dying* is a logical consequence of the four premises *Socrates is a man*, *All men are mortal*, *No mortal lives forever*, and *Everyone who will eventually die sometimes worries about it.* A proof of this conclusion might pass through the following intermediate steps. First we note that from the first two premises it follows that Socrates is mortal. From this intermediate conclusion and the third premise (that no mortal lives forever), it follows that Socrates will eventually die. But this, along with the fourth premise, gives us the desired conclusion, that Socrates sometimes worries about dying.

A proof that S follows from some premises P, Q, ... may be quite long and complicated. But each step in the proof is supposed to provide absolutely incontrovertible evidence that the intermediate conclusion follows from things already established. Here, the logician's standards of rigor are extreme. It is not enough to show that each step in a purported proof almost certainly follows from the ones that come before. That may be good enough for getting around in our daily life, but it is not good enough if our aim is to demonstrate that S *must* be true provided P, Q, ... are all true.

There is a practical reason for this demand for rigor. In ordinary life, we frequently reason in a step-by-step fashion, without requiring absolute certainty at each step. For most purposes, this is fine, since our everyday "proofs" generally traverse only a small number of intermediate conclusions. But in many types of reasoning, this is not the case. Think of what you did in high school geometry. First you started with a small number of axioms that stated the basic premises of Euclidean geometry. You then began to prove conclusions, called theorems, from these axioms. As you went on to prove more interesting theorems, your proofs would cite earlier theorems. These earlier theorems were treated as intermediate conclusions in justifying the new results. What this means is that the complete proofs of the later theorems really include the proofs of the earlier theorems that they presuppose. Thus, if they were written out in full, they would contain hundreds or perhaps thousands of steps. Now suppose we only insisted that each step show with probability .99 that the conclusion follows from the premises. Then each step in such a proof would be a pretty good bet, but given a long enough proof, the proof would carry virtually no weight at all about the truth of the conclusion.

This demand for certainty becomes even more important in proofs done by computers. Nowadays, theorems are sometimes proven by computers, and the proofs can be millions of steps long. If we allowed even the slightest uncertainty in the individual steps, then this uncertainty would multiply until the alleged "proof" made the truth of the conclusion no more likely than its falsity.

Remember

A proof of a statement S from premises P, Q, ... is a step-by-step demonstration which shows that S *must* be true in any circumstances in which the premises P, Q, ... are all true.

In each of the Chapters 2–7, we will devote two sections to discussing methods of proof. In the first of these sections, we will discuss the main methods of proof used in mathematics, emphasizing the more important methods, such as proof by contradiction and conditional proof. In the second of these sections, we will "formalize" the methods presented in the preceding section.

The difference between an informal proof and a formal proof is not one of rigor, but of style. An informal proof of the sort used by mathematicians is every bit as rigorous as a formal proof. But it is stated in English and is usually more free-wheeling, leaving out the more obvious steps. A formal proof, by contrast, employs a fixed stock of rules and a highly stylized method of presentation. For example, the simple little argument that we presented earlier will, in our formal system, take the following form:

1. Cube(c)	
2. c = b	
3. Cube(b)	Ind Id: 1, 2

We will explain later the conventions used in this formal proof.

In the course of this book you will learn how to give both informal and formal proofs. We do not want to give the impression that formal proofs are somehow better than informal proofs. On the contrary, for purposes of proving things for ourselves, or communicating proofs to others, informal methods are usually preferable. Formal proofs come into their own in two ways. One is that they display the logical structure of a proof in a form that can be mechanically checked. There are advantages to this, if you are a logic teacher grading lots of homework, a computer, or not inclined to think for some other reason. The other is that they allow us to prove things about provability itself, such as Gödel's Completeness Theorem and Incompleteness Theorems, discussed in the final section of this book.

There are not many methods of proof that can be discussed when only atomic sentences are available. Still, there are some, to which we now turn.

Proofs involving the identity symbol

We have already seen one example of an important method of proof. If we can prove, from whatever our premises happen to be, that b = c then we know that anything we can prove about b holds of c. After all, b *is* c. This simple observation sometimes goes by the fancy name of the indiscernibility of identicals.

This principle is used repeatedly in mathematics. Thus, for example, the following derivation uses the principle in conjunction with the well-known algebraic identity $(x-1)(x+1) = x^2 - 1$:

$$x^2 > x^2 - 1$$

so

$$x^2 > (x-1)(x+1)$$

We are all familiar with reasoning that uses such substitutions repeatedly.

Another principle, so simple that one often overlooks it, is the so called reflexivity of identity. It tells us that any sentence of the form a = a can be validly inferred from whatever premises are at hand, or from no premises at all. This is because of a basic assumption we make in FOL, namely, that names always refer to one and only one object. This is not true about English, as we have noted before. But it is in FOL, which means that in a proof one can always take any name a that is in use and assert a = a, if it suits your purpose for some reason. (As a matter of fact, it is seldom of much use.)

Another principle, a bit more useful, is that of the symmetry of identity. It allows us to conclude b = a from a = b. Actually, if we wanted, we could derive this as a consequence of our first two principles, as follows.

> **Proof:** Suppose that a = b. We know that a = a, by the reflexivity of identity. Now replace the first use of the name a in a = a by b using the indiscernibility of identicals. We come up with b = a, as desired.

The previous paragraph is an example of an informal proof. We will see how to formalize it in the next section.

A third principle about identity that bears noting is its so-called transitivity. If a = b and b = c are both true, then so is a = c. This is so obvious that there is no particular need to prove it, but it can be proved using the indiscernibility of identicals. (See Problem 18.)

Proofs using other atomic sentences

Sometimes there will be logical dependencies among the relation symbols in a first-order language, dependencies similar to those just discussed involving the identity symbol. This is the case, for example, in the language used in Tarski's World. When this is so, proofs may exploit such relationships. For example, the sentence Larger(a, c) is a consequence of Larger(a, b) and Larger(b, c). This is because the larger-than relation, like the identity relation, is transitive. It is because of this that any world where the latter two sentences are true will also be one where the first is true. Similar examples are given in the problems.

An example of this sort that is used frequently in mathematics involves the transitivity of the less-than relation. One frequently encounters proofs written in the following form:

$$\begin{aligned} k_1 &< k_2 \\ k_2 &< k_3 \\ k_3 &< k_4 \end{aligned}$$

so

$$k_1 < k_4$$

This contains two implicit uses of the transitivity of $<$.

One final remark, before doing some problems. When we say that S is a logical consequence of premises P, Q, ..., we do not insist that each of the premises really play an essential role. So, for example, if S is a logical consequence of P then it is also a logical consequence of P and Q. This follows immediately from the definition of logical consequence. But it has a corollary for our notion of proof. We do not insist that each of the premises in a proof is actually used in the proof.

Exercises and Problems

Problem 18. Give a proof of a = c from the premises a = b and b = c using only the indiscernibility of identicals.

Problem$_\diamond$ 19. Given the meaning of the atomic predicates in Tarski's World, which of the following are logical consequences of the stated premises? If the conclusion is not a consequence of the premises, build a world in which the premises are true and the conclusion false. (Name such worlds World 2.19.x, where x is the number of the argument below.)

1. Premise: LeftOf(a, b); Conclusion: RightOf(b, a)
2. Premise: LeftOf(a, b), b = c; Conclusion: RightOf(c, a)
3. Premises: LeftOf(a, b), RightOf(c, a); Conclusion: LeftOf(b, c)

4. Premises: BackOf(a, b), FrontOf(a, c); Conclusion: FrontOf(b, c)
5. Premises: Between(b, a, c), LeftOf(a, c); Conclusion: LeftOf(a, b)
   [Hint: This one is tricky; see Exercise 1.]

Problem 20. Consider the following sentences.

1. *Folly was Claire's disk at 2 pm.*
2. *Silly was Max's disk at 2 pm.*
3. *Silly was not Claire's disk at 2 pm.*

Does (3) follow from (1) and (2)? Does (2) follow from (1) and (3)? Does (1) follow from (2) and (3)? In each case, if your answer is *no*, describe a possible circumstance in which the premises are true and the conclusion false.

2.9 Formal proofs

In this section we will begin introducing a particular system for presenting formal proofs, what is known as a "deductive system." There are many different styles of deductive systems. The system we present in the first three parts of the book, which we will call $\mathcal{F}$, is what is known as a "Fitch-style" system, named after the logician Frederic Fitch, who first introduced this type of system. We will study a very different deductive system in Part IV, one known as the resolution method, which is of considerable importance in computer science.

In general, a proof in the system $\mathcal{F}$ from premises P, Q, R, of a conclusion S, takes the following form:

```
| P
| Q
| R
|---
| ⋮
| S
```

There are two graphical devices to notice here, the vertical and horizontal lines. The vertical line draws our attention to the fact that we have a single purported proof consisting of several steps. The horizontal line, known as the "Fitch bar," indicates a division between the claims that are assumed and those that allegedly follow from them. Thus the fact that P, Q, and R are above the line shows that these are the premises of our proof, while the fact that S is below the line shows that S is supposed to follow. In an actual proof, the three dots would be replaced by whatever intermediate conclusions, if any, are needed to derive S from the premises. Also, in giving an actual formal proof, we will number the steps, for ease of reference in justifying later steps.

We have already given one example of a formal proof in the system $\mathcal{F}$. For another example, here is a formalization of our earlier proof of the symmetry of identity.

1. a = b	
2. a = a	Refl =
3. b = a	Ind Id: 2, 1

In the right hand margin of this proof you will find the logical justification for each step below the Fitch bar. These are applications of rules we are about to introduce. The numbers at the right of step 3 show that this step follows from steps 2 and 1 by means of the rule cited, the rule Ind Id. (The reason for ordering them this way will be explained when we describe Ind Id.)

The first rule we use in the above proof is the Reflexivity of Identity. This rule allows you to assert, for any name n in use in the proof, the sentence n = n. You are allowed to do this at any step in the proof, and need not cite any earlier step as justification. We will abbreviate our statement of this rule in the following way:

Reflexivity of Identity (Refl =):

▷ | n = n

We have used an additional graphical device in stating this rule. This is the symbol ▷. We will use it in stating rules to indicate which step is being licensed by the rule. In this example there is only one step, but in other examples there will be several steps.

The second rule of $\mathcal{F}$ is the Indiscernibility of Identicals. It tells us that if we have proven a sentence containing the name n (which we indicate by writing P(n)) and a sentence of the form n = m, then we are justified in asserting any sentence which results from P(n) by replacing some or all of the occurrences of n by m. To apply this rule it does not matter at what point in the proof we have derived P(n) and n = m, or even what order. All that matters is that they occur before the conclusion P(m).

Indiscernibility of Identicals (Ind Id):

| P(n)
| ⋮
| n = m
| ⋮
▷ | P(m)

When we apply this rule, it does not matter which of P(n) and n = m occurs first in the proof, as long as they both appear before P(m), the inferred step. In justifying the step, we cite the name of the rule, followed by the steps in which P(n) and n = m occur, in the order in which they occur.

Now that we have the rules before us, let us give another example of a formal proof. Suppose we were asked to give a formal proof of Likes(b, a) from premises Likes(a, a) and b = a. We would begin by writing down the premises and the conclusion, leaving space in between for our proof.

```
| 1. Likes(a, a)
| 2. b = a
|---
|  ⋮
|
| Likes(b, a)
```

We then fill in the missing ingredients of the proof, step by step. The finished proof looks like this. Make sure you understand why all the steps are there.

```
| 1. Likes(a, a)
| 2. b = a
|---
| 3. b = b                      Refl =
| 4. a = b                      Ind Id: 3, 2
| 5. Likes(b, a)                Ind Id: 1, 4
```

We could also introduce into the system $\mathcal{F}$ rules justified by the meanings of other predicates besides =, like Between and Smaller. For example, we could introduce a formal rule of the following sort:

Bidirectionality of Between:

```
  | Between(a, b, c)
  |  ⋮
▷ | Between(a, c, b)
```

We don't do this because there are just too many such rules. We could state them for a few predicates, but certainly not all of the predicates you will encounter in first-order languages.

We will, however, add one rule that is not technically necessary, but that will make some proofs look more natural. This rule is called Reiteration, and simply allows you to repeat an earlier step, if you so desire.

Reiteration (Reit):

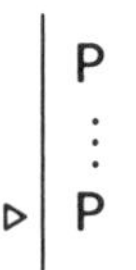

To use the Reiteration rule, just repeat the sentence in question and, on the right, write "Reit: x," where x is the number of the earlier occurrence of the sentence.

Reiteration is obviously a valid rule of inference, since any sentence is a logical consequence of itself. The reason for having the rule will become clear as proofs in the system $\mathcal{F}$ become more complicated. For now, let's just say that it is like remarking, in the course of giving an informal proof, "we have already shown that P." This is often a helpful reminder to the person reading the proof.

Exercises and Problems

Problem 21. Formalize the proof you gave in Problem 18.

Problem 22. Give a formal proof of the sentence Likes(a, d) from premises Likes(a, b), b = c, and c = d.

Problem 23. Give a formal proof of the sentence Between(c, d, e) from the premises Between(a, d, b), a = c, and e = b.

Problem 24. The predicate Smaller in the language of Tarski's World is transitive. That is to say, Smaller(a, c) follows from Smaller(a, b) and Smaller(b, c).

1. State this in the form of a Fitch-style rule.
2. Use this rule to give a formal proof of Smaller(a, d) from Smaller(a, b), Smaller(c, d), and b = c.

Problem 25. State Fitch-style rules expressing the relation between RightOf and LeftOf. Use your rules to give a formal proof of RightOf(a, c) from LeftOf(b, a) and b = c.

2.10 Alternative notation

As we said before, FOL is like a family of languages. But, as if that were not enough diversity, even the very same first-order language comes in a variety of dialects. Indeed, almost no two logic books use exactly the same notational conventions in writing first-order sentences. For this reason, it is important to have some familiarity with the different dialects—the different notational conventions—and to be able to translate smoothly between them. At the end of each chapter, we discuss common notational differences that you are likely to encounter.

Some notational differences, though not many, occur even at the level of atomic sentences. For example, some authors insist on putting parentheses around atomic sentences whose binary predicates are in infix position. So (a = b) is used rather than a = b. By contrast, some authors omit parentheses surrounding the argument positions (and the commas between them) when the predicate is in prefix position. These authors use Rab instead of R(a, b). We have opted for the latter simply because we use predicates made up of several letters, and the parentheses make it clear where the predicate ends and the arguments begin: Cubed is not nearly as perspicuous as Cube(d).

What is important in these choices is that sentences should be unambiguous and easy to read. Typically, the first aim requires parentheses to be used in one way or another, while the second suggests using no more than is necessary.

3

Conjunctions, Disjunctions, and Negations

To form complex claims, FOL provides us with connectives and quantifiers. In this chapter we take up the three simplest connectives: conjunction, disjunction, and negation. These are three of the *truth-functional* connectives. Being truth-functional means that the truth value of a complex sentence built up using these connectives depends on nothing more than the truth values of the simpler sentences from which it is built.

We can explain the meaning of a truth-functional connective in a variety of ways. Perhaps the easiest is by constructing a *truth table*, a table that shows how the truth value of a sentence formed with the connective depends on the truth values of the sentence's immediate parts. We will give such tables for each of the connectives we introduce. A more interesting way, however, is by means of a game, sometimes called the Henkin-Hintikka game, after the logicians Leon Henkin and Jaakko Hintikka. We simply call it the game.

Imagine that two people, say Max and Claire, disagree about the truth value of a complex sentence. Max claims it is true, Claire claims it is false. The idea of the game is that the two players keep challenging each other to justify their claims in terms of simpler claims, until finally their disagreement is reduced to a simple atomic claim, one involving an atomic sentence. At that point they can simply examine the world to see whether the atomic claim is true—at least in the case of claims about the sorts of worlds we find in Tarski's World. These successive challenges can be thought of as a game where one player will win, the other will lose. The legal moves at any stage depend on the form of the sentence. We will explain them below. The one who can ultimately justify his or her claims is the winner.

When you play this game in Tarski's World, the computer takes the opposite side, even if it knows you are right. If you are mistaken in your initial assessment, the computer will be sure to win the game. If you are right, though, the computer plugs away, hoping you will blunder. If you do, the computer will win the game. We will use the game rules as a second way of explaining the meanings of the truth-functional connectives.

3.1 Negation symbol (¬)

This symbol is used to express negation in our language, the notion we commonly express in English using terms like *not*, *it is not the case that*, *non-* and *un-*. In first-order logic, we always apply this symbol to the front of a sentence to be negated, while in English there is a much more subtle system for expressing negative claims. For example, the English sentences *John isn't home* and *It is not the case that John is home* have the same first-order translation:

¬Home(John)

This sentence is true if and only if Home(John) isn't true, that is, just in case John isn't home.

In English, we generally avoid double negatives—negatives inside other negatives. For example, the sentence *It doesn't make no difference* is problematic. If someone says it, they usually mean that it doesn't make any difference. In other words, the second negative just functions as an intensifier of some sort. On the other hand, this sentence could be used to mean just what it says, that it does not make *no* difference, it makes *some* difference.

FOL is much more systematic. You can put a negation symbol in front of any sentence whatsoever, and it always negates it, no matter how many other negation symbols the sentence already contains. For example, the sentence

¬¬Home(John)

negates the sentence

¬Home(John)

and so is true if John is home.

The negation symbol, then, can apply to complex sentences as well as to atomic sentences. We will say that a sentence is a *literal* if it is either atomic or the negation of an atomic sentence. This notion of a literal will be useful later on.

We will abbreviate negated identity claims, such as ¬(b = c), using ≠, as in b ≠ c. The symbol ≠ is available on the Tarski's World Keyboard.

Semantics and the game rule for negation

Given any sentence P of FOL (atomic or complex), there is another sentence ¬P. This sentence is true if and only if P is false. This can be expressed in terms of the following truth table.

P	¬P
TRUE	FALSE
FALSE	TRUE

The game rule for negation is very simple, since you never have to *do* anything. Once you commit yourself to the truth of ¬P this is tantamount to committing yourself to the falsity of P. Similarly, if you commit yourself to the falsity of ¬P, this is the same as committing yourself to the truth of P. So in either case Tarski's World simply replaces your commitment about the more complex sentence by the opposite commitment about the simpler sentence.

Exercises and Problems

Exercise$_\diamond$ 1. Open Wittgenstein's World. Start a new sentence list and write the following sentence. Then see if it is true.

¬¬¬¬¬Between(c, b, d)

Now play the game, committing yourself to the truth of the sentence. What happens to the number of negation symbols as the game proceeds? What happens to your commitment?

Problem 2. Let P be a true sentence, and let Q be formed by putting some number of negation symbols in front of P. Show that if you put an even number of negation symbols, then Q is true, but that if you put an odd number, then Q is false. [Hint: A complete proof of this simple fact would require what is known as "mathematical induction." If you are familiar with proof by induction, then go ahead and give a proof. If you are not, just convince yourself that you understand why this is true.]

Now assume that P is atomic. Show that no matter how many negation symbols Q has, it will have the same truth value as a literal, namely either the literal P or the literal ¬P.

3.2 Conjunction symbol (∧)

This symbol is used to express conjunction in our language, the notion we normally express in English using terms like *and*, *moreover*, and *but*. In first-order logic, this connective is always placed between two sentences, whereas in English we can also conjoin nouns, verbs, and other parts of speech. For example, the English sentences *John and Mary are home* and *John is home and Mary is home* both have the same first-order translation:

Home(John) ∧ Home(Mary)

This sentence is true just in case John is home and Mary is home.

Some uses of the English *and* are not accurately mirrored by the FOL conjunction symbol. For example, suppose we are talking about an evening when Max and Claire were together. If we were to say *Max went home and Claire went to sleep*, our assertion would carry with it a temporal implication, namely that Max went home *before* Claire went to sleep. Similarly, if we were to reverse the order and assert *Claire went to sleep and Max went home* it would suggest a very different sort of situation. By contrast, no such implication, implicit or explicit, is intended when we use the FOL conjunction ∧. The sentence

WentHome(Max) ∧ FellAsleep(Claire)

is true in exactly the same circumstances as

FellAsleep(Claire) ∧ WentHome(Max)

Semantics and the game rule for ∧

Just as with negation, we can put complex sentences as well as simple ones together with ∧. A sentence P ∧ Q is true if and only if both P and Q are true. Thus P ∧ Q is false if either or both of P or Q is false. This can be summarized by the following truth table.

P	Q	P ∧ Q
TRUE	TRUE	TRUE
TRUE	FALSE	FALSE
FALSE	TRUE	FALSE
FALSE	FALSE	FALSE

The Tarski's World game is more interesting for conjunctions. The way the game proceeds depends on whether you have committed to true or to false. If you commit to the truth of P ∧ Q then you have implicitly committed yourself to the truth of each of P and Q. Thus Tarski's World gets to choose either one of these simpler sentences and hold you to the truth of it. (Which one will Tarski's World choose? If one of them is false, it will choose that one. If both are true, or both are false, it will choose at random.)

If you commit to the falsity of P ∧ Q then you are claiming that at least one of P or Q is false. Tarski's World will ask you to choose one of the two and thereby explicitly commit yourself to its being false.

Exercises and Problems

Exercise$_\diamond$ 3. Start a new sentence list and open Wittgenstein's World. Write the following sentences in the list, saving them as Sentences 3.3.

1. Tet(f) ∧ Small(f)
2. Tet(f) ∧ Large(f)
3. Tet(f) ∧ ¬Small(f)
4. Tet(f) ∧ ¬Large(f)
5. ¬Tet(f) ∧ ¬Small(f)
6. ¬Tet(f) ∧ ¬Large(f)

7. ¬(Tet(f) ∧ Small(f))
8. ¬(Tet(f) ∧ Large(f))
9. ¬(¬Tet(f) ∧ ¬Small(f))
10. ¬(¬Tet(f) ∧ ¬Large(f))

Once you have written these sentences, decide which you think are true. Record your evaluations, to help you remember. Then go through and use Tarski's World to evaluate your assessments. Whenever you are wrong, play the game to see where you went wrong. If you are never wrong, then play the game a couple times anyway, knowing that you should win. Next, change the size or shape of f, predict how this will affect the truth values of these sentences, and see if your prediction is right. What is the maximum number of these sentences that you can get to be true in a single world?

3.3 Disjunction symbol (∨)

This symbol is used to express disjunction in our language, the notion we express in English using *or*. In first-order logic, this connective, like the conjunction sign, is always placed between two sentences, whereas in English we can also disjoin nouns, verbs, and other parts of speech. For example, the English sentences *John or Mary is home* and *John is home or Mary is home* both have the same first-order translation:

Home(John) ∨ Home(Mary)

Although the English *or* is sometimes used in an "exclusive" sense, to say that *exactly* one of the two disjoined sentences is true, the first-order logic ∨ is always given an "inclusive" interpretation: it means that at least one and possibly both of the two disjoined sentences is true. Thus, our sample sentence is true if John is home but Mary is not, if Mary is home but John is not, or if both John and Mary are home.

If we wanted to express the exclusive sense of *or* in the above example, we could do it as follows:

[Home(John) ∨ Home(Mary)] ∧ ¬[Home(John) ∧ Home(Mary)]

As you can see, this sentence says that John or Mary is home, but they are not both home.

Many students are tempted to say that the English expression *either ... or* expresses exclusive disjunction. While this is sometimes the case (and indeed the simple *or* is often used exclusively), it isn't always. For example, suppose Claire and Max are playing in the other room and the sound of a fight suddenly breaks out. If we say *Either Max hit Claire or Claire hit Max*, we would not be wrong if each had hit the other. So this would be translated as

$$\text{Hit(Max, Claire)} \lor \text{Hit(Claire, Max)}$$

We will see later that the expression *either* sometimes plays a different logical function.

Another important English expression that we can capture without introducing additional symbols is *neither... nor*. Thus *Neither John nor Mary is at home* would be expressed as:

$$\neg(\text{Home(John)} \lor \text{Home(Mary)})$$

This says that it's not the case that at least one of them is at home, i.e., that neither of them is home.

Semantics and the game rule for ∨

Given any two sentences P and Q of FOL, atomic or not, we can combine them using ∨ to form a new sentence P ∨ Q. The sentence P ∨ Q is true if at least one of P or Q is true. Otherwise it is false. Here is the truth table.

P	Q	P ∨ Q
TRUE	TRUE	TRUE
TRUE	FALSE	TRUE
FALSE	TRUE	TRUE
FALSE	FALSE	FALSE

The game rules for ∨ are the "duals" of those for ∧. Thus, if you commit yourself to the truth of P ∨ Q, then Tarski's World will make you live up to this by committing yourself to the truth of one or the other. If you commit yourself to the falsity of P ∨ Q, then you are implicitly committing yourself to the falsity of each, so Tarski's World will choose one and hold you to the commitment that it is false. (Tarski's World will, of course, try to win by picking a true one, if it can.)

Exercises and Problems

Exercise◇ 4. Open Wittgenstein's World and the sentence file Sentences 3.3 that you created in Exercise 3. Edit the sentence list by replacing ∧ by ∨ throughout. Once you have changed these sentences, decide which you think are true. Again, record your evaluations to help you remember them. Then go through and use Tarski's World to evaluate your assessment. Whenever you are wrong, play the game to see where you went wrong. If you are never wrong, then play the game anyway a couple times, knowing that you should win. As in Exercise 3, what is the maximum number of sentences you can make true by changing the size or shape (or both) of f? How does this compare with the corresponding result in Exercise 3?

TABLE 2
Game rules for ∧, ∨, and ¬

FORM	YOUR COMMITMENT	PLAYER TO MOVE	GOAL
P ∨ Q	TRUE FALSE	you Tarski's World	Choose one of P, Q that is true.
P ∧ Q	TRUE FALSE	Tarski's World you	Choose one of P, Q that is false.
¬P	either	—	Replace ¬P by P and switch commitment.

Remarks about the game

We summarize the game rules for the three connectives, ¬, ∧, and ∨, in Table 2. The first column indicates the form of the sentence in question, and the second indicates your current commitment, TRUE or FALSE. Which player moves depends on this commitment, as shown in the third column. The goal of that player's move is indicated in the final column. Notice that although the player to move depends on the commitment, the goal of that move does not depend on the commitment. This is because when you are committed to TRUE, Tarski's World is committed to FALSE, and vice versa.

There is one somewhat subtle point that needs to be made about our way of describing the game. We have said, for example, that when you are committed to the truth of a disjunction P ∨ Q, you are committed to the truth of one of the disjuncts. This of course is true, but does not mean you necessarily know which of P or Q is true. For example, if you have a sentence of the form P ∨ ¬P, then you know that it is true, no matter how the world is. But if P is quite complex, or if you have imperfect information about the world, you may not know which of P or ¬P is true. Suppose P is a sentence like Fermat's Conjecture or perhaps *There is a whale swimming below the Golden Gate Bridge right now.* In such cases you would be willing to commit to the truth of the disjunction without knowing just how to play the game and win. You know that there is a winning strategy for the game, but just don't know what it is.

Since there is a moral imperative to live up to one's commitments, the use of the term "commitment" in describing the game is a bit misleading.

You are perfectly justified in asserting the truth of $\mathsf{P} \lor \neg\mathsf{P}$, even if you do not happen to know your winning strategy for playing the game. Indeed, it would be foolish to claim that the sentence is *not* true. But if you do claim that $\mathsf{P} \lor \neg\mathsf{P}$ is true, and then play the game, you will be asked to say which of P or ¬P you think is true. Tarski's World has been designed so you can always get complete information about the world, and so always live up to such commitments. This is illustrated in Problem 14 later in the chapter.

3.4 Ambiguity and parentheses

When we first described FOL, we stressed the lack of ambiguity of this language as opposed to ordinary languages. For example, English allows us to say things like *Max is home or Claire is home and Carl is happy.* This sentence can be understood in two quite different ways. One reading claims that either Claire is home and Carl is happy, or Max is home. On this reading, the sentence would be true if Max was home, even if Carl was unhappy. The other reading claims both that Max or Claire is home and that Carl is happy.

FOL avoids this sort of ambiguity by requiring the use of parentheses, much the way they are used in algebra. So, for example, FOL would not have one sentence corresponding to the ambiguous English sentence, but two:

$$\mathsf{Home(Max)} \lor (\mathsf{Home(Claire)} \land \mathsf{Happy(Carl)})$$
$$(\mathsf{Home(Max)} \lor \mathsf{Home(Claire)}) \land \mathsf{Happy(Carl)}$$

The parentheses in the first indicate that it is a disjunction, whose second disjunct is itself a conjunction. In the second, they indicate that the sentence is a conjunction whose first conjunct is a disjunction. As a result, the truth conditions for the two are quite different.

Parentheses are also used to indicate the "scope" of a negation symbol when it appears in a complex sentence. So, for example, the two sentences

$$\neg\mathsf{Home(Claire)} \land \mathsf{Home(Max)}$$
$$\neg(\mathsf{Home(Claire)} \land \mathsf{Home(Max)})$$

mean quite different things. The first is a conjunction of literals, the first of which says Claire is not home, the second of which says that Max is home. By contrast, the second sentence is a negation of a sentence which itself is a conjunction: it says that they are not both home. You have already encountered this use of parentheses in earlier exercises.

Many logic books require that you always put parentheses around any pair of sentences joined by a binary connective (such as ∧ or ∨). So for example, the following sentence, though not ambiguous, would not be allowed:

Home(Max) ∧ Home(Claire) ∧ Happy(Carl)

These books would require that we write one of the following:

((Home(Max) ∧ Home(Claire)) ∧ Happy(Carl))
(Home(Max) ∧ (Home(Claire) ∧ Happy(Carl)))

Tarski's World is not so fussy, in a couple of ways. First of all, it allows us to conjoin any number of sentences without using parentheses, since the result is not ambiguous, and similarly for disjunctions. Second, it allows us to leave off the outermost parentheses, since they serve no useful purpose. It also allows you to add extra parentheses (or brackets or braces) if you want to for the sake of readability. For the most part, all Tarski's World requires is that your expression be unambiguous.

Let's summarize what we have learned so far about the three connectives we have introduced.

Remember

1. ¬P is true if and only if P is not true.
2. P ∧ Q is true if and only if P is true and Q is true.
3. P ∨ Q is true if and only if P is true or Q is true (or both are true).
4. Parentheses must be used whenever ambiguity would result from their omission. In practice, what this means is that conjunctions and disjunctions must be "wrapped" in parentheses whenever combined by means of some other connective.

To really master a new language, you have to use it, not just read about it. The exercises and problems that follow are intended to let you do just that.

Exercises and Problems

Exercise$_\diamond$ 5. (Basic propositional logic) In this exercise you are asked to evaluate some sentences built up from atomic sentences using the three connectives ∧, ∨, ¬. Open Boole's Sentences, and evaluate the sentences in Wittgenstein's World. (If you changed the size or shape of f while doing Exercises 3 and 4, make sure that you change it back to a large tetrahedron.) If you make a mistake, play the game to see where you have gone wrong. Don't go from one sentence to the next until you understand why it has the truth value it does. Do you see the importance of parentheses?

Problem$_\diamond$ **6.** (Building a world) Open Schröder's Sentences. Build a single world where all the sentences in this file are true. As you work through the sentences, you will find yourself successively modifying the world. Whenever you make a change in the world, be careful that you don't make one of your earlier sentences false. When you are finished, verify that all the sentences are really true. Save your world as World 3.6.

Problem$_\diamond$ **7.** (DeMorgan Equivalences) Open the file DeMorgan's Sentences. Construct a world where all the *odd* numbered sentences are true. Notice that no matter how you do this, the even numbered sentences also come out true. Save this world as World 3.7.1. Next build a world where all the odd numbered sentences are *false*. Notice that no matter how you do it, the even numbered sentences also come out false. Save this world as World 3.7.2. Can you explain why this is so?

3.5 Logical equivalence

There are two very general and important facts behind the scene in Problem 7, known as *DeMorgan's laws*. To express these laws, we first introduce some terminology. We say that two sentences P and Q in the same first-order language are *logically equivalent* if they are true in exactly the same circumstances (and so false in the same circumstances). We will write this as:

$$P \Leftrightarrow Q$$

If P and Q are logically equivalent sentences in the language of Tarski's World, then of course the same worlds will make them true. If there is a world in which one is true and the other false, then they are not equivalent.

As an example, it is obvious that for any sentence P, the sentences P and $\neg\neg P$ are logically equivalent. This is known as the law of *double negation*. Even simpler examples are the so-called *association rules*. So, for example, $P \land (Q \land R)$ is logically equivalent to $(P \land Q) \land R$, which in turn is equivalent to $P \land Q \land R$. The same holds with $\land$ replaced by $\lor$ throughout. We will never mention the association laws again, since they are so simple.

DeMorgan's laws are more significant. One tells us that the negation of a conjunction, $\neg(P \land Q)$, is logically equivalent to the disjunction of the negations of the original conjuncts: $\neg P \lor \neg Q$. The other tells us that the negation of a disjunction, $\neg(P \lor Q)$, is logically equivalent to the conjunction of the negations of the original disjuncts: $\neg P \land \neg Q$. These are simple consequences of the meanings of the connectives involved.

When two sentences are logically equivalent, then each is a logical consequence of the other. As a result, in giving an informal proof, one can always go from any sentence to one that is logically equivalent to it. This fact makes observations like the DeMorgan laws quite useful in giving informal proofs.

Remember

(Double negation and DeMorgan's Laws) For any sentences P and Q:

1. Double negation: $\neg\neg P \Leftrightarrow P$
2. DeMorgan: $\neg(P \wedge Q) \Leftrightarrow (\neg P \vee \neg Q)$
3. DeMorgan: $\neg(P \vee Q) \Leftrightarrow (\neg P \wedge \neg Q)$

Using these three laws, one can take any sentence built up with $\wedge, \vee$, and $\neg$, and transform it into one where $\neg$ applies only to atomic sentences. Another way of expressing this is that any sentence built out of atomic sentences using the three connectives $\wedge, \vee$, and $\neg$ is logically equivalent to one built from literals using just $\wedge$ and $\vee$. To obtain such a sentence, one simply drives the $\neg$ in, switching $\wedge$ to $\vee$, $\vee$ to $\wedge$, and canceling any pair of $\neg$'s that are right next to each other, not separated by any parentheses. Such a sentence is said to be in *negation normal form*. Here is an example of a derivation of the negation normal form of a sentence. We use A, B, and C to stand for any *atomic* sentences of the language.

$$\begin{aligned} \neg((A \vee B) \wedge \neg C) &\Leftrightarrow \neg(A \vee B) \vee \neg\neg C \\ &\Leftrightarrow (\neg A \wedge \neg B) \vee \neg\neg C \\ &\Leftrightarrow (\neg A \wedge \neg B) \vee C \end{aligned}$$

In reading and giving derivations of this sort, remember that the symbol $\Leftrightarrow$ is not itself a symbol of the first-order language. Rather, it is a shorthand way of saying that two sentences are logically equivalent.

We end this section with a list of some additional logical equivalences and some exercises relating to them.

1. (Idempotence of $\wedge$) If a conjunction has a repeated conjunct then the conjunction is logically equivalent to the result of removing all but one occurrence of that conjunct. For example,

$$P \wedge Q \wedge P \Leftrightarrow P \wedge Q$$

2. (Idempotence of $\vee$) If a disjunction has a repeated disjunct then the disjunction is logically equivalent to the result of removing all

but one occurrence of that disjunct. For example,

$$P \lor Q \lor P \Leftrightarrow P \lor Q$$

3. (Commutativity of $\land$) Any rearrangement of the conjuncts of a sentence of FOL is logically equivalent to the original. For example,

$$P \land Q \land R \Leftrightarrow Q \land P \land R$$

4. (Commutativity of $\lor$) Any rearrangement of the disjuncts of a sentence of FOL is logically equivalent to the original. For example,

$$P \lor Q \lor R \Leftrightarrow Q \lor P \lor R$$

Here is an example where we use some of these laws to show that the first sentence in the following list is logically equivalent to the last. Once again (as in what follows) we use A, B, and C to stand for arbitrary atomic sentences of FOL. Thus the result is in negation normal form.

$$\begin{aligned}
(A \lor B) \land C \land (\neg(\neg B \land \neg A) \lor B) &\Leftrightarrow (A \lor B) \land C \land ((\neg\neg B \lor \neg\neg A) \lor B) \\
&\Leftrightarrow (A \lor B) \land C \land ((B \lor A) \lor B) \\
&\Leftrightarrow (A \lor B) \land C \land (B \lor A \lor B) \\
&\Leftrightarrow (A \lor B) \land C \land (B \lor A) \\
&\Leftrightarrow (A \lor B) \land C \land (A \lor B) \\
&\Leftrightarrow (A \lor B) \land C
\end{aligned}$$

In this list the first step is justified by one of the DeMorgan laws. The second step involves two applications of Double Negation. In the next step we use association to remove the unnecessary parentheses. In the fourth step, we use Idempotence of $\lor$. The next to the last step uses Commutativity of $\lor$, while the final step uses Idempotence of $\land$.

Exercises and Problems

Problem$_\diamond$ 8. (Negation normal form) Open Turing's Sentences. You will note that every other sentence is blank. In each space write the negation normal form of the sentence above it. Then build any world where all of the names are in use. If you have gotten the negation normal forms correct, each even numbered sentence will have the same truth value in your world as the sentence above it. Verify that this is so in your world. Save the sentences as Sentences 3.8.

Problem 9. (Practice using commutativity and idempotence) Use the above laws, plus the removal of any unnecessary parentheses, to simplify the following expressions as much as possible.

1. $(A \land B) \land A$
2. $(B \land (A \land B \land C))$
3. $(A \lor B) \lor (C \land D) \lor A$
4. $(\neg A \lor B) \lor (B \lor C)$
5. $(A \land B) \lor C \lor (B \land A) \lor A$

Problem 10. (Negation normal form) Use Double Negation and DeMorgan's laws to put the following sentences into negation normal form.

1. ¬(Home(Carl) ∧ ¬Home(Claire))
2. ¬[Happy(Max) ∧ (¬Likes(Carl, Claire) ∨ ¬Likes(Claire, Carl))]
3. ¬¬¬[(Home(Max) ∨ Home(Carl)) ∧ (Happy(Max) ∨ Happy(Carl))]

3.6 Translation

An important skill that you will want to master is that of translating from English to FOL, and vice versa. But before you can do that, you need to know how to express yourself in both languages. The problems below are designed to help you learn these related skills.

How do we know if a translation is correct? Intuitively, a correct translation is a sentence with the same meaning as the one being translated. But what is the meaning? FOL finesses this question, settling for truth conditions. What we require of a correct translation in FOL is that it be true in the same circumstances as the original sentence. For sentences of Tarski's World, this boils down to being true in the very same worlds.

Note that it is not sufficient that the two sentences have the same truth value in some *particular* world. If that were so, then any true sentence of English could be translated by any true sentence of FOL. So, for example, if Claire and Max are both at home, we could translate *Max is at home* by means of Home(Claire). No, having the same actual truth value is not enough. They have to have the same truth values in all circumstances.

Remember

In order for an English sentence and an FOL sentence to have the same meaning, it is necessary that they have the same truth values in all circumstances.

In general, this is all we require of translations into and out of FOL. Thus, given an English sentence *S* and a good FOL translation of it, say S, any other sentence S′ that is logically equivalent to S will also count as an acceptable translation of it. (Why?) But there is a matter of style. Some good translations are better than others. You want sentences that are easy to understand. But you also want to keep the FOL connectives close to the English, if possible. For example, a good translation of *It is not true that Claire and Max are both at home* would be given by

¬(Home(Claire) ∧ Home(Max))

This is logically equivalent to the following negation normal form sentence, so we have to count it too as a good translation.

$$\neg\text{Home(Claire)} \lor \neg\text{Home(Max)}$$

But there is a clear stylistic sense in which the first is a better translation, since it conforms more closely to the form of the original. There are no hard and fast rules for determining which among several logically equivalent sentences is the best translation of a given sentence.

In Problem 12, sentences 1, 8, and 10, you will discover an important function that the English phrases *either... or* and *both... and* sometimes play. *Either* helps disambiguate the following *or* by indicating how far to the left its scope extends; similarly *both* indicates how far to the left the following *and* extends. For example, *Either Max is home and Claire is home or Carl is happy* is unambiguous, whereas it would be ambiguous without the *either*. What it means is that

$$[\text{Home(Max)} \land \text{Home(Claire)}] \lor \text{Happy(Carl)}$$

In other words, *either* and *both* can sometimes act as left parentheses act in FOL. The same list of sentences demonstrates many other uses of *either* and *both.*

Exercises and Problems

Problem$_\diamond$ 11. (Describing a simple world) Open Boole's World. Start a new sentence file, named Sentences 3.11, where you will describe some features of this world. Check each of your sentences to see that it is indeed a sentence and that it is true in this world.

1. Notice that f (the large dodecahedron in the back) is not in front of a. Use your first sentence to say this.
2. Notice that f is to the right of a and to the left of b. Use your second sentence to say this.
3. Use your third sentence to say that f is either in back of or smaller than a.
4. Express the fact that both e and d are between c and a.
5. Note that neither e nor d is larger than c. Use your fifth sentence to say this.
6. Notice that e is neither larger than nor smaller than d. Use your sixth sentence to say this.
7. Notice that c is smaller than a but larger than e. State this fact.
8. Note that c is in front of f; moreover, it is smaller than f. Use your eighth sentence to state these things.

Now let's change the world so that none of the above mentioned facts hold. We can do this as follows. First move f to the front right corner of the grid. (Be careful not to drop it off the edge. You might find it easier to make the move from the 2-D view. If you accidentally drop it, just open Boole's World again.) Then move e to the back left corner of the grid and make it large. Now none of the facts hold; if your answers to 1–8 are correct, all of the sentences should now be false. Verify that they are. If any are still true, can you figure out where you went wrong?

Problem$_\diamond$ 12. (Some translations) Tarski's World provides you with a very useful way to check whether your translation of a given English sentence is correct. If it is correct, then it will always have the same truth value as the English sentence, no matter what world the two are evaluated in. So when you are in doubt about one of your translations, simply build some worlds where the English sentence is true, others where it is false, and check to see that your translation has the right truth values in these worlds. You should use this technique frequently in all of the translation exercises.

Start a new sentence file, and use it to enter translations of the following English sentences into first-order logic. You will only need to use the connectives $\wedge, \vee$, and $\neg$.

1. *Either **a** is small or both **c** and **d** are large.*
2. ***d** and **e** are both in back of **b**.*
3. ***d** and **e** are both in back of **b** and larger than it.*
4. *Both **d** and **c** are cubes; moreover neither of them is small.*
5. *Neither **e** nor **a** is to the right of **c** and to the left of **b**.*
6. *Either **e** is not large or it is in back of **a**.*
7. ***c** is neither between **a** and **b**, nor in front of either of them.*
8. *Either both **a** and **e** are tetrahedra or both **a** and **f** are.*
9. *Neither **d** nor **c** is in front of either **c** or **b**.*
10. ***c** is either between **d** and **f** or smaller than both of them.*

Save your list of sentences as Sentences 3.12.

Exercise$_\diamond$ 13. (Checking your translations) Open Wittgenstein's World. Notice that all of the English sentences from Exercise 12 are true in this world. Thus, if your translations are accurate, they will also be true in this world. Check to see that they are. If you made any mistakes, go back and fix them. But as we have stressed, even if one of your sentences comes out true in Wittgenstein's World, it does not mean that it is a proper translation of the corresponding English sentence. All you know for sure is that your translation and the original sentence have the same truth value in this particular world. If the translation is correct, it will have the same truth value as the English sentence in *every* world. Thus, to have a better test of your translations, we will examine them in a number of worlds, to see if they have the same truth values as their English counterparts in all of these worlds.

Let's start by making modifications to Wittgenstein's World. Make all the large or medium objects small, and the small objects large. With these changes in the world, the English sentences 1, 3, 4, and 10 become false, while the rest remain true. Verify that the same holds for your translations. If not, correct your translations. Next, rotate your modified Wittgenstein's World 90° to the right. Now sentences 5, 6, 8, and 9 should be the only true ones that remain.

Let's check your translations in another world. Open Boole's World. The only English sentence that is true in this world is sentence 6. Verify that all of your translations except 6 are false. If not, correct your translations.

Now modify Boole's World by exchanging the positions of b and c. With this change, the English sentences 2, 5, 6, and 7 come out true, while the rest are false. Check that the same is true of your translations.

Problem$_{\diamond}$ 14. Here is a problem that illustrates the remarks we made earlier about sometimes being able to tell that a sentence is true, without knowing how to win a play of the game.

Open Kleene's World and Kleene's Sentences. Some objects are hidden behind other objects, thus making it impossible to assess the truth of some of the sentences. Each of the six names a, b, c, d, e, and f are in use, naming some object. Now even though you cannot see all the objects, some of the sentences in the list can be evaluated with just the information at hand. Assess the truth of each claim, if you can, without recourse to the 2-D view. Then play the game. If your initial commitment is right, but you lose the game, back up and play over again. Then go through and add comments to each sentence explaining whether you can assess its truth in the world as shown, and why. Finally, display the 2-D view and check your work. We have annotated the first sentence for you to give you the idea. When you are done, save your annotated sentences as Sentences 3.14.

Problem 15. Using the names and predicates presented in Table 1 on page 23, translate the following into FOL.

1. *Max is a student, not a disk.*
2. *Claire erased Folly at 2 pm and then gave it to Max.*
3. *Folly belonged to either Max or Claire at 2:05 pm.*
4. *Neither Max nor Claire erased Folly at 2 pm or at 2:05 pm.*
5. *2:00 pm is between 1:55 pm and 2:05 pm.*
6. *When Max gave Folly to Claire at 2 pm, it wasn't blank, but it was five minutes later.*

Problem 16. Referring again to Table 1, page 23, translate the following into natural, colloquial English.

1. Student(Claire) ∧ ¬Student(Max)
2. Disk(Silly) ∧ ¬Owned(Max, Silly, 2:00)
3. Owned(Claire, Silly, 2:00) ∨ Owned(Claire, Folly, 2:00)
4. ¬(Erased(Max, Silly, 2:00) ∧ Erased(Max, Folly, 2:00))
5. ((Gave(Max, Silly, Claire, 2:00) ∧ Blank(Silly, 2:00)) ∨ (Gave(Max, Folly, Claire, 2:00) ∧ Blank(Folly, 2:00))) ∧ Angry(Claire, 2:05)

Problem* 17. Translate the following into FOL, introducing names, predicates, and function symbols as needed. Explain the meaning of each predicate and function symbol, unless it is completely obvious.

1. *AIDS is less contagious than influenza, but more deadly.*
2. *Abe fooled Stephen on Sunday, but not on Monday.*
3. *Dan or George admires Al and Bill.*
4. *Daisy is a jolly miller, and lives on the River Dee.*
5. *Polonius's eldest child was neither a borrower nor a lender.*

3.7 Satisfiability and logical truth

There are two important notions that are closely related to the notion of logical equivalence. The first is that of a *satisfiable* sentence, or a satisfiable set of sentences. Intuitively, a sentence is satisfiable if it could be true, at least on logical grounds. There might be some other reason, say physical, why it might not be possible, but there are no logical reasons why it couldn't be true. The way it is usually put is that there is some logically possible circumstance in which the sentence is true.

A set of sentences is satisfiable if there is some possible circumstance in which all of the sentences in the set are simultaneously true. Thus, it is not enough that each sentence in the list be satisfiable on its own. For example, each of the following three sentences is satisfiable. Indeed, any two of them are satisfiable at once. But the whole set is not satisfiable.

Happy(Claire) ∨ Happy(Max)
¬Happy(Claire)
¬Happy(Max)

A sentence of the Tarski's World version of FOL is satisfiable if we can build a world in which it is true. It turns out, though, that there are satisfiable sentences that can't be made true in the worlds you can build with Tarski's World. For example, the sentence

¬(Tet(b) ∨ Cube(b) ∨ Dodec(b))

is satisfiable, say, if b were a sphere. You can't build such a world with Tarski's World, but that is not logic's fault. It's the fault of the program designers. In earlier problems you have already shown that various sets of sentences were satisfiable, by building worlds that make them true.

The second notion is that of a *logically true* sentence. A sentence is logically true if it must be true no matter what the world is like. Thus, for example, the following are each logically true.

$$\mathsf{Home(Claire)} \vee \neg\mathsf{Home(Claire)}$$
$$\neg(\mathsf{Happy(Carl)} \wedge \neg\mathsf{Happy(Carl)})$$
$$\neg[(\mathsf{Happy(Carl)} \vee \mathsf{Happy(Max)}) \wedge \neg\mathsf{Happy(Carl)} \wedge \neg\mathsf{Happy(Max)}]$$

There is a simple relationship between these two notions. One way of expressing it is to say that a sentence P is logically true if and only if its negation ¬P is not satisfiable. Checking that this is so is simply a matter of unraveling the two definitions we have just given. Another way to express it is that a sentence Q is satisfiable if and only if ¬Q is not logically true.

There is a simple method for determining when a sentence built up from atomic sentences by means of $\wedge, \vee$, and $\neg$ is satisfiable or not, logically true or not. The method involves building a truth table for the sentence in question. Before we explain the method, we must describe how to build truth tables for complex sentences.

Suppose we have a complex sentence S whose atomic sentences are $A_1, \ldots, A_n$. To build a truth table for S, one writes the atomic sentences $A_1, \ldots, A_n$ across the top of the page, with the sentence S to their right. It is customary to draw a double line separating the atomic sentences from S. Your truth table will have one row for every possible way of assigning TRUE and FALSE to the atomic sentences. Since there are two possible assignments to each atomic sentence, there will be 2^n rows. Thus if $n = 1$ there will be two rows, if $n = 2$ there will be four rows, if $n = 3$ there will be eight rows, and so forth. It is customary to make the leftmost column have the top half of the rows marked TRUE, the second half false. The next row splits each of these, marking the first and third quarters of the rows with TRUE, the second and fourth quarters with FALSE, and so on. This will result with the last column having TRUE and FALSE alternating down the column.

Let's look at an example with three atomic sentences.

$$(\mathsf{Cube(a)} \wedge \mathsf{Cube(b)}) \vee \neg\mathsf{Cube(c)}$$

In order to make our table easier to read, we will abbreviate the atomic sentences by A, B, and C. We will later fill in the column under the complex sentence.

A	B	C	(A ∧ B) ∨ ¬C
T	T	T	
T	T	F	
T	F	T	
T	F	F	
F	T	T	
F	T	F	
F	F	T	
F	F	F	

In a truth table, the columns under the atomic sentences are called "reference columns." Once the reference columns have been filled in, we are ready to fill in the remainder of the table. To do this, we construct columns of T's and F's beneath each connective of the target sentence S. These columns are filled in one by one, using the truth tables for the various connectives. We start by working on connectives that apply only to atomic sentences. Once this is done, we work on connectives that apply to sentences whose main connective has already had its column filled in. We continue this process until the main connective of S has had its column filled in. This is the column that shows how the truth of S depends on the truth of its atomic parts.

Let's carry out the first step of this process for the above table. Since two of the connectives in the target sentence apply to atomic sentences, we can fill in two columns using the truth tables for ∧ and ¬ given earlier.

A	B	C	(A ∧ B)	∨	¬C
T	T	T	T		F
T	T	F	T		T
T	F	T	F		F
T	F	F	F		T
F	T	T	F		F
F	T	F	F		T
F	F	T	F		F
F	F	F	F		T

This leaves only one connective, the main connective of the sentence. We fill it in by referring to the two columns just completed, and using the truth table for ∨.

A	B	C	(A ∧ B)	∨	¬C
T	T	T	T	T	F
T	T	F	T	T	T
T	F	T	F	F	F
T	F	F	F	T	T
F	T	T	F	F	F
F	T	F	F	T	T
F	F	T	F	F	F
F	F	F	F	T	T

It needs to be noted that some of the rows in a truth table may be spurious, in that they may not represent genuine possibilities. This is not the case in the above example, but imagine that the atomic sentence A had been a = a. In that case the whole bottom half of the table would have been spurious, since a = a could not possibly be false. Likewise, if A had been the sentence Tet(c), then any row that assigned TRUE to both A and C would have been spurious, since c cannot be both a tetrahedron and a cube. So if we want to know whether our sentence is logically true, we will have to check to see whether any of the rows are spurious. If so, then these rows should be ignored.

There is, in general, no simple mechanical method for recognizing whether a row in a truth table is spurious, or whether it represents a genuine possibility. This will depend on the meanings of the atomic sentences involved. For this reason, it is useful to introduce a technical notion that applies to those sentences which come out true in *every* row, spurious or not. Such sentences are called *tautological sentences* or simply *tautologies.* All tautological sentences are logically true, of course, since being a tautology is a more stringent condition. Some logically true sentences are not tautological, however. The simplest example of such a sentence is a = a.

The distinction between tautologies and logical truths is usually sloughed over in treatments of propositional logic. Authors simply arrange things so that no spurious rows arise in their truth tables. This requires two things. First, if you have an atomic sentence A, then it must be possible for each of A and ¬A to be true. Thus, the language cannot contain a sentence like a = a. Second, there must be no logical relationships between distinct atomic sentences, such as the relationship that holds between Larger(a, b), Larger(b, c), and Larger(a, c). If these conditions hold in a given language, then we will say that its atomic sentences are *logically independent.* In most languages, the atomic sentences are not logically independent.

Now that we know how to build truth tables, we can describe the method for determining whether a sentence is satisfiable, logically true, or tautological.

Given: A sentence S built up from atomic sentences $A_1, \ldots, A_n$.

Goal: To determine whether S is satisfiable, logically true, or tautological.

Method:

1. Construct a truth table for the sentence S.
2. S is a tautology if and only if it is true in every row of the table.
3. Eliminate any spurious rows from the table.
4. S is satisfiable if and only if there is at least one remaining row that assigns TRUE to S.
5. S is logically true if and only if every remaining row assigns TRUE to S.

In our sample truth-table, there are no spurious rows, so we see that the sentence is satisfiable, but not logically true.

There is, in fact, another method for showing that a sentence is logically true, one that uses the technique of proofs. If you can prove a sentence using no premises whatsoever, then the sentence is logically true. Later in the chapter we will give you some more methods for giving proofs. Using these, you will be able to prove that sentences are logically true without constructing their truth tables. When we add quantifiers to our language, the truth table method will no longer apply, but the method of proof will still work.

Remember

Let S be a sentence of FOL built up from atomic sentences by means of truth functional connectives alone. A truth table for S shows how the truth of S depends on the truth of its atomic parts.

1. S is a tautology if and only if every row assigns TRUE to S.
2. S is satisfiable if and only if there is at least one non-spurious row of the truth table that assigns TRUE to S.
3. S is logically true if and only if every non-spurious row assigns TRUE to S.

Exercises and Problems

Problem 18. Assume that A, B, and C are atomic sentences. Which of the following are tautologies?

1. $(A \land B) \lor (\neg A \lor \neg B)$

2. $(A \wedge B) \vee (A \wedge \neg B)$
3. $\neg(A \wedge B) \vee C$
4. $(A \vee B) \vee \neg(A \vee (B \wedge C))$

Problem 19. Suppose that the sentences A, B, and C in Problem 18 are not logically independent. Determine whether the sentences (1)-(4) are satisfiable or logically true under each of the following assumptions:

1. A is necessarily true. (For example, A is $b = b$.)
2. A is necessarily false. (For example, A is Smaller(a, a).)
3. C is a logical consequence of A and B, that is, whenever both A and B are true, so is C. (For example, A is Larger(a, b), B is Larger(b, c), and C is Larger(a, c).)

Problem 20. Let us call a sentence S *tt-satisfiable* if there is at least one row of its truth table (spurious or not) that assigns S TRUE. In other words, tt-satisfiable sentences are to satisfiable sentences as tautologies are to logical truths.

1. Explain why every satisfiable sentence is tt-satisfiable.
2. Give an example from Tarski's World of a sentence that is tt-satisfiable but not satisfiable.
3. Explain why every tt-satisfiable sentence is satisfiable if the atomic sentences of the language in question are logically independent.
4. Explain why a sentence S is tt-satisfiable if and only if ¬S is not a tautology.

Problem 21. Let us reexamine the truth table for $(A \wedge B) \vee \neg C$ constructed on page 54, but this time assuming there are logical dependencies among the atomic sentences. For each of the following, indicate which rows of the truth table (if any) would need to be eliminated as spurious.

1. A is the wff $b = b$ with B and C as originally specified.
2. A is the wff Tet(b) with B and C as originally specified.
3. A, B, and C are the wffs Cube(a), Tet(a), and Dodec(a), respectively. (Note that all of these can be false, though not in a world build with Tarski's World.)
4. C is the wff $b \neq b$ but A and B are as in the original.
5. A, B, and C are the wffs Larger(a, b), Larger(b, c), and Larger(c, a), respectively.

Problem 22. Assume that A, B, C, and D are logically independent atomic sentences. Construct truth tables for each of the following sentences:

1. $\neg(B \wedge \neg(C \vee B))$

2. A ∨ ¬(B ∨ ¬(C ∧ A))
3. ¬[(¬A ∨ ¬(B ∧ C)) ∨ (A ∧ B)]
4. ¬[(¬A ∨ B) ∧ ¬(C ∧ D)]

Indicate which are satisfiable and which are logically true.

Problem 23. Write out truth tables for the following. Which rows, if any, are spurious? Indicate them by crossing them out.

1. Happy(Carl) ∧ ¬(Home(Carl) ∨ Home(Claire))
2. Happy(Carl) ∨ ¬(Carl = Carl ∧ Claire = Claire)
3. [Happy(Carl) ∨ ¬Happy(Carl)] ∨ Home(Max)
4. [Happy(Carl) ∨ Unhappy(Carl)] ∨ Home(Max)
   Note: Sometimes Carl is neither happy nor unhappy, as when he is asleep.
5. Likes(Carl, Scruffy) ∨ Dislikes(Carl, Scruffy)

Problem$_\diamond$ 24. (Logical dependencies) Open Weiner's Sentences. Fill in a table of the following sort.

Sentence	Tarski's World	Truth assignment
1		
2		
3		
⋮		

In the first column, put *yes* if it is possible to make the sentence true by building a world, and *no* otherwise. If your answer is *yes* for a sentence, then construct such a world and save it as World 3.24.x, where x is the number of the sentence in question. In the second column, put *yes* if it is possible to come up with an assignment of truth values to the atomic sentences which makes the sentence true. If your answer is *yes*, then write out your truth assignment. Indicate whether or not the assignment is spurious. Note that sometimes there will be rows in your truth table that are not spurious, and yet do not correspond to any world you can build using Tarski's World.

Problem* 25. (Logical consequence) Truth tables can be used to check for logical consequence as well as for logical truth. Suppose, for example, you want to know whether Q is a logical consequence of P. Create a truth table where you list all the atomic sentences that appear in either P or Q. Following the double vertical line, form a column for P. Following a second double vertical line, put another column for Q. Fill in both these columns using the method described in this section. Then if every non-spurious row in which P is true is also one in which Q is true, then Q is a logical consequence of P. If *every row* (spurious or not) in which

P is true is one in which Q is true, then Q is said to be a *tautological consequence* of P. Use this method answer the following questions:

1. Is Cube(b) a tautological consequence of (Cube(a) ∨ Cube(b)) ∧ Tet(a)?
2. Is Cube(b) a logical consequence of (Cube(a) ∨ Cube(b)) ∧ Tet(a)?
3. Is Cube(b) a tautological consequence of (Cube(a) ∨ Cube(b)) ∧ ¬Cube(a)?
4. Is Cube(b) a logical consequence of (Cube(a)∨Cube(b))∧¬Cube(a)?
5. Is Dodec(c) a tautological consequence of Dodec(b) ∧ c = b?
6. Is Dodec(c) a logical consequence of Dodec(b) ∧ c = b?
7. Is Dodec(c) is a logical consequence of Dodec(b) ∧ e = c?

Problem* 26. Suppose that S is a tautology, with atomic sentences A, B, and C. Suppose that we replace all occurrences of A by another sentence P, possibly complex. Explain why the resulting sentence is still a tautology. This is expressed by saying that substitution preserves tautologicality. Explain why substitution does not preserve logical truth. Give an example.

3.8 Methods of proof involving ¬, ∧, and ∨

Each new connective we introduce into our language gives rise to legitimate patterns of inference. Some of these are extremely simple, involving simple entailments between pairs of sentences, like the entailment from P ∧ Q to P. These we will refer to as *valid inference steps.* More interesting, however, are new *methods of proof* that are allowed by the new expressions. We will discuss these at greater length.

Valid steps

Probably the simplest example of a valid step in a proof is to follow any sentence P by any other sentence Q that is known to be logically equivalent to it. For this reason, it is handy to have certain logical equivalences at your fingertips. We have already discussed some of the more interesting equivalences involving ¬, ∧, and ∨. We will discuss two other important equivalences in Section 3.10.

A simple instance of this involves use of the principle of Double Negation, in either direction. For example, given the sentence ¬¬LeftOf(a, b), you can infer LeftOf(a, b), and vice versa.

In a similar vein, we can use logical equivalences to replace *parts* of sentences by logically equivalent parts. Thus if we have proven some sentence containing the sentence ¬¬LeftOf(a, b) we can replace this in the original by LeftOf(a, b). Sentences that differ only in logically equivalent parts are themselves logically equivalent.

A related principle that can be useful is the following: If you know that a sentence Q is a logical truth, then you may assert Q at any point in your proof. We have already seen this principle at work in Chapter 2, when we discussed the reflexivity of identity, the principle that allowed us to assert a sentence of the form (b = b) at any point in a proof. It also allows us to assert other simple logical truths, like $\mathsf{P} \vee \neg\mathsf{P}$, at any point in a proof.

In addition, there are three other simple but important valid steps that need to be mentioned, two involving $\wedge$ and one involving $\vee$.

Suppose we have managed to prove a conjunction, say $\mathsf{P} \wedge \mathsf{Q}$. Then it clearly follows that P and it clearly follows that Q. More generally, we are justified in inferring, from a conjunction of any number of sentences, any one of its conjuncts. Why? Because there is no way for the conjunction to be true without each conjunct being true. These steps are sometimes called "conjunction elimination" or "simplification," when they are presented in the context of a formal system of deduction. When they are used in actual mathematical reasoning, however, they usually go by without comment, since they are so obvious.

Only slightly more interesting is the converse. Suppose we have managed to prove a sentence P and another sentence, Q, from the same premises. Then, given the meaning of $\wedge$, we are clearly entitled to infer the conjunction $\mathsf{P} \wedge \mathsf{Q}$. For if both P and Q are true, so is the conjunction. More generally, if we want to prove a conjunction of a bunch of sentences, we may do so by proving each conjunct separately. In a formal system of deduction, steps of this sort are sometimes called "conjunction introduction" or just "conjunction." Once again, in real life reasoning, these steps are too simple to warrant mention. In our informal proofs, we will seldom mention them explicitly.

Finally, let us look at one valid step involving $\vee$. It is a simple step, but it is one that strikes students as peculiar.

Suppose, for example, that you have proven Cube(b). Then you can conclude Cube(a) $\vee$ Cube(b) $\vee$ Cube(c), if you should want to for some reason. More generally, if you have proven some sentence P then you can infer any disjunction that has P as one of its disjuncts. After all, if P is true, so is any such disjunction.

What strikes newcomers to logic as peculiar about such a step is that using it amounts to throwing away information. Why in the world would you want to conclude $\mathsf{P} \vee \mathsf{Q}$ when you already know the more informative claim P? But as we will see, this step is actually quite useful when combined with some of the methods of proof to be discussed later. Still, in mathematical proofs, it generally goes by unnoticed. In formal systems, it is dubbed "disjunction introduction," or (rather unfortunately) "addition."

These are by no means the only valid steps, or even the only important ones. They are, however, enough to give you the feeling for the simple nature of such steps. We present some others in the following problem.

Exercises and Problems

Problem 27. (Additional valid steps) In the following we list a number of inference steps, only some of which are valid. For each, determine whether it is valid. If it is, explain why it is valid, appealing to the truth tables for the connectives involved. If it is not, give an example of how the step could be used to get from true premises to a false conclusion.

1. From $\mathsf{P} \vee \mathsf{Q}$ and $\neg\mathsf{P}$ infer Q.
2. From $\mathsf{P} \vee \mathsf{Q}$ and Q infer $\neg\mathsf{P}$.
3. From $\neg(\mathsf{P} \vee \mathsf{Q})$ infer $\neg\mathsf{P}$.
4. From $\neg(\mathsf{P} \wedge \mathsf{Q})$ and P infer $\neg\mathsf{Q}$.
5. From $\neg(\mathsf{P} \wedge \mathsf{Q})$ infer $\neg\mathsf{P}$.
6. From $\mathsf{P} \wedge \mathsf{Q}$ and $\neg\mathsf{P}$ infer Q.

Method of proof by cases

In addition to simple, valid proof steps, the symbols we have introduced give rise to two significant methods of proof. Unlike the above, these methods are often explicitly applied in mathematical reasoning. They are the method of proof by cases and the method of proof by contradiction. In formal systems of inference, these sometimes go by the names of disjunction elimination and negation introduction, respectively.

Consider the following well-known piece of reasoning. It proves that there are irrational[1] numbers b and c such that b^c is rational.

> **Proof:** To show that there are irrational numbers b and c such that b^c is rational, we consider the number $\sqrt{2}^{\sqrt{2}}$. We note that this is either rational or irrational. If it is rational, then we have found our b and c; namely, we take $b = c = \sqrt{2}$. If it is irrational, though, then we take $b = \sqrt{2}^{\sqrt{2}}$ and $c = \sqrt{2}$ and compute b^c:
>
> $$\begin{aligned} b^c &= (\sqrt{2}^{\sqrt{2}})^{\sqrt{2}} \\ &= \sqrt{2}^{(\sqrt{2}\cdot\sqrt{2})} \\ &= \sqrt{2}^2 \\ &= 2 \end{aligned}$$

[1] A number is rational if it can be expressed as a fraction n/m, for integers n and m. Otherwise it is irrational. Thus 2 is rational ($2 = 2/1$) but $\sqrt{2}$ is irrational, as we will soon show.

Thus, we see that in this case, too, b^c is rational. So, whether $\sqrt{2}^{\sqrt{2}}$ is rational or irrational, we know that there are irrational numbers b and c such that b^c is rational.

Notice the general structure of this argument. We begin with a desired goal that we want to prove, say S, and a disjunction we already know, say P ∨ Q. We then show two things: that S follows if we assume that P is the case, and that S follows if we assume that Q is the case. Since we know that one of these must hold, we then conclude that S must be the case. This pattern of reasoning is called "disjunction elimination" or "proof by cases."

There is no need to break into just two cases, as we have in our example. If at any stage in a proof we have some disjunction containing n disjuncts, say $\mathsf{P}_1 \vee \ldots \vee \mathsf{P}_n$, then we can break into n cases. In the first we assume P_1, in the second P_2, and so forth for each disjunct. If we are able to prove our desired result S in each of these cases, we are justified in concluding that S holds.

Let us look at an even simpler example of proof by cases. Suppose we want to prove that Small(c) is a logical consequence of

(Cube(c) ∧ Small(c)) ∨ (Tet(c) ∧ Small(c)).

This is, of course, rather obvious, but the proof involves breaking into cases, as you will notice if you think carefully about how you recognize this. For the record, here is how we would write out the proof.

Proof: We are given

(Cube(c) ∧ Small(c)) ∨ (Tet(c) ∧ Small(c))

as a premise. We will break into two cases, corresponding to the two disjuncts. First, assume that Cube(c) ∧ Small(c) holds. But then (by conjunction elimination, which we really shouldn't even mention) we have Small(c). But likewise, if we assume Tet(c) ∧ Small(c), then it follows that Small(c). So, in either case, we have Small(c), as desired.

Let's look at one more example, one that shows how the odd step of disjunction introduction can be used fruitfully with proof by cases. Suppose we know that either Max is home and Carl is happy, or Claire is home and Scruffy is happy, i.e.,

(Home(Max) ∧ Happy(Carl)) ∨ (Home(Claire) ∧ Happy(Scruffy))

We want to prove that either Carl or Scruffy is happy, that is,

Happy(Carl) ∨ Happy(Scruffy)

A rather pedantic, step-by-step proof would look like this:

Proof: Assume the disjunction:

(Home(Max) ∧ Happy(Carl)) ∨ (Home(Claire) ∧ Happy(Scruffy))

Then either

(Home(Max) ∧ Happy(Carl))

or

(Home(Claire) ∧ Happy(Scruffy)).

If the first alternative holds, then Happy(Carl), and so we have

Happy(Carl) ∨ Happy(Scruffy)

by disjunction introduction. Similarly, if the second alternative holds, we have Happy(Scruffy), and so

Happy(Carl) ∨ Happy(Scruffy)

So, in either case, we have our desired conclusion. Thus our conclusion follows by proof by cases.

Remember

(Proof by cases) To prove S from $P_1 \vee \ldots \vee P_n$, prove S from each of $P_1, \ldots, P_n$.

Exercises and Problems

Problem 28. In our proof that there are irrational numbers b and c where b^c is rational, one of our steps was to assert that $\sqrt{2}^{\sqrt{2}}$ is either rational or irrational. What justifies the introduction of this claim into our proof?

Method of proof by contradiction

One of the most important methods of proof is known as proof by contradiction. (It is also called *Reductio ad absurdum* and negation introduction.) The basic idea is this. Suppose you want to prove a negative sentence, say ¬S, from some premises, say P, Q, ... One way to do this is by temporarily assuming S and showing that a contradiction follows from this assumption. If you can show this, then you are entitled to conclude that ¬S is a logical consequence of the original premises. Why? Because your proof of the contradiction shows that S, P, Q, ... cannot all be true simultaneously. Hence if P, Q, ... are true in any set of circumstances, then S must be false in those circumstances.

Let us look at a simple example of this method of proof. Assume Cube(c) ∨ Dodec(c) and Tet(b). Let us prove ¬(b = c).

Proof: In order to prove $\neg(b = c)$, we assume $b = c$ and attempt to get a contradiction. From our first premise we know that either Cube(c) or Dodec(c). If the first is the case, then we get Cube(b) by indiscernibility of identicals, which contradicts Tet(b). But similarly, if the second is the case, we get Dodec(b) which contradicts Tet(b). So neither case is possible, and we have a contradiction. Thus our initial assumption that $b = c$ must be wrong. So proof by contradiction gives us our desired conclusion, $\neg(b = c)$. (Notice that this argument also uses the method of proof by cases.)

Let us now give a more interesting and famous example of this method of proof. The Greeks were shocked to discover that the square root of 2 could not be expressed as a fraction, or, as we would put it, is irrational. The proof of this fact proceeds via contradiction. Before we go through the proof, let's review some simple numerical facts that were well known to the Greeks. The first is that any rational number can be expressed as a fraction p/q where at least one of p and q is odd. (If not, keep dividing both the numerator and denominator by 2 until one of them is odd.) The other fact follows from the observation that when you square an odd number, you always get an odd number. So if n^2 is an even number, then so is n. And from this, we see that if n^2 is even, it must be divisible by 4.

Now for the proof that $\sqrt{2}$ is irrational.

Proof: With an eye toward getting a contradiction, we will suppose that $\sqrt{2}$ is rational. Thus it can be expressed in the form $\sqrt{2} = p/q$, where at least one of p and q is odd. Since $p/q = \sqrt{2}$ we have

$$\frac{p^2}{q^2} = 2$$

Multiplying both sides by q^2, we get $p^2 = 2q^2$. But this shows that p^2 is an even number. As we noted before, this allows us to conclude that p is even and that p^2 is divisible by 4. Looking again at the equation $p^2 = 2q^2$, we see that if p^2 is divisible by 4, then q^2 must be divisible by 2. In which case, q is even as well. So both p and q are even, contradicting the fact that at least one of them was odd. Thus our assumption that $\sqrt{2}$ is rational led us to a contradiction, and so we conclude that it is irrational.

In order to apply the method of proof by contradiction, it is important that you understand what a contradiction is, since that is what you need to prove. Intuitively, a contradiction is some claim that cannot possibly be true, or some set of claims which cannot all be true simultaneously. Examples are a sentence Q and its negation ¬Q, a sentence of the form $b \neq b$, or a pair of sentences like Cube(c) and Tet(c). We can in fact define the notion of a contradictory set of sentences as any set which is not satisfiable.

You can also use proof by contradiction to prove a sentence S that does not begin with a negation. In this case, you would begin by assuming ¬S, obtain a contradiction, and then conclude that ¬¬S is the case, which of course is equivalent to S.

> **Remember**
>
> (Proof by contradiction) To prove ¬S using this method, assume S and prove some sort of contradiction.

Inconsistent premises

What are the logical consequences of an inconsistent set of premises? If you look back at our definitions, you will see that every sentence S is a consequence of such a set. After all, if the premises are inconsistent, then there is no way that they can all be true. Thus, in a vacuous way, every set of circumstances that does make them true also makes S true.

For example, suppose you are given the following premises:

1. Home(Max) ∨ Home(Claire)
2. ¬Home(Max)
3. ¬Home(Claire)

Then every sentence in the same first-order language counts as a logical consequence of these premises. But unlike a normal valid argument, an argument with an inconsistent set of premises is not worth much. After all, the reason we are interested in logical consequence is because of its relation to truth. If the premises can't possibly be true, then even knowing that the argument is valid gives us no clue as to the truth or falsity of the conclusion.

Exercises and Problems

In these problems we ask you to prove various results. Your proofs should be phrased in complete, well-formed English sentences, making use of first-order sentences as convenient, much in the style we have used above. Whenever you use one of the two methods just discussed, say so. You don't have to be explicit about the use of simple proof steps like conjunction elimination. By the way, there is typically more than one way to prove a given result.

Problem 29. Prove Happy(Carl) from the following three premises:

1. $\neg$(Home(Max) $\wedge$ Home(Claire))
2. Home(Max) $\vee$ Happy(Carl)
3. Home(Claire) $\vee$ Happy(Carl)

Problem$_\diamond$ 30. Assume the following as premises.

1. LeftOf(a, b) $\vee$ RightOf(a, b)
2. BackOf(a, b) $\vee$ $\neg$LeftOf(a, b)
3. FrontOf(b, a) $\vee$ $\neg$RightOf(a, b)
4. Large(c)

Is BackOf(a, b) a logical consequence of these premises? If so, give a proof. If not, build a world in which all the premises are true but the conclusion is false, and save it as World 3.30.

Problem$_\diamond$ 31. Assume the same four premises as in Problem 30. Is $\neg$Between(c, b, a) a logical consequence of these premises? If so, give a proof. If not, build a world in which the premises are true and the conclusion is false, and save it as World 3.31.

Problem$_\diamond$ 32. Assume the following as premises.

1. Cube(a) $\vee$ Tet(a) $\vee$ Large(a)
2. $\neg$Cube(a) $\vee$ a = b $\vee$ Large(a)
3. $\neg$Large(a) $\vee$ a = c
4. $\neg$(c = c $\wedge$ Tet(a))

Is (a = b) $\vee$ (a = c) a logical consequence of these premises? If so, give a proof. If not, build a world in which all the premise are true but the conclusion is false and save it as World 3.32.

Problem$_\diamond$ 33. Assume the same four premises as in Problem 32. Is $\neg$(Large(a) $\vee$ Tet(a)) a logical consequence of these premises? If so, give a proof. If not, build a world in which the premises are true and the conclusion is false and save it as World 3.33.

Problem 34. Translate the following into FOL and then give a proof that the fourth sentence follows from the first three.

1. Max or Claire is at home but either Scruffy or Carl is unhappy.
2. Either Max is not home or Carl is happy.
3. Either Claire is not home or Scruffy is unhappy.
4. Scruffy is unhappy.

Problem 35. Assume the following are about a world of the kind that arises in Tarski's World. Give a proof that c is not small, from the following premises:

1. b is a tetrahedron.
2. c is a cube.
3. Either c is larger than b or else they are identical.

Problem 36. Consider the following sentences.

1. *Folly was Claire's disk at 2 pm or at 2:05 pm.*
2. *Folly was Max's disk at 2 pm.*
3. *Folly was Claire's disk at 2:05 pm.*

Does (3) follow from (1) and (2)? Does (2) follow from (1) and (3)? Does (1) follow from (2) and (3)? In each case, give either a proof of consequence, or describe a situation that makes the premises true and the conclusion false. You may assume that only one person may own a disk at any given time.

Problem 37. In this problem we ask you to prove three simple facts about the natural numbers. We do not expect you to phrase the proofs in FOL. You will need to appeal to basic facts of arithmetic plus the definitions of even and odd number.[2] This is OK, but make these appeals explicit. You will find proof by contradiction helpful.

1. Assume n^2 is odd. Prove that n is odd.
2. Assume $n + m$ is odd. Prove that $n \times m$ is even.
3. Assume n^2 is divisible by 3. Prove that n^2 is divisible by 9.

Problem* 38. A good way to make sure you understand a proof is to try to generalize it. Prove that $\sqrt{3}$ is irrational. [Hint: You will need to figure out some facts about divisibility by 3 that parallel the facts we used about even and odd, for example, the third result in Problem 37.] Can you generalize these two results?

3.9 Formal proofs

In this section we add rules of inference to the formal system $\mathcal{F}$ that correspond to the valid principles of reasoning introduced in the previous section. For each connective, we will introduce two rules, one that allows us to prove statements containing the symbol, and one that allows us to prove things from statements containing the symbol. The former are called "introduction" rules since they allow us to introduce these symbols into proofs. By contrast, the latter are called "elimination" rules.

[2] A number n is even if it is divisible by 2, that is, if there is a number k such that $n = 2k$; it is odd if it is not even.

Conjunction rules

The simplest principles to formalize are those that involve the conjunction symbol $\wedge$. These are the rules of conjunction elimination and conjunction introduction.

The rule of conjunction elimination allows you to assert any conjunct P_i of a conjunctive sentence $P_1 \wedge \ldots \wedge P_i \wedge \ldots \wedge P_n$ that you have already derived in the proof. (P_i can, by the way, be any conjunct, including the first or the last.) You justify the new step by citing the number of the step containing the conjunction. We abbreviate this rule with the following schema:

Conjunction Elimination ($\wedge$ Elim):

$$\begin{array}{l|l} & P_1 \wedge \ldots \wedge P_i \wedge \ldots \wedge P_n \\ & \vdots \\ \triangleright & P_i \end{array}$$

The corresponding introduction rule, conjunction introduction, allows you to assert a conjunction $P_1 \wedge \ldots \wedge P_n$ provided you have already established each of its constituent conjuncts P_1 through P_n. We will symbolize this rule in the following way:

Conjunction Introduction ($\wedge$ Intro):

$$\begin{array}{l|l} & P_1 \\ & \Downarrow \\ & P_n \\ & \vdots \\ \triangleright & P_1 \wedge \ldots \wedge P_n \end{array}$$

In this rule, we have used the notation:

$$\begin{array}{c} P_1 \\ \Downarrow \\ P_n \end{array}$$

to indicate that each of P_1 through P_n must appear in the proof before you can assert their conjunction. The order in which they appear does not matter, and they do not have to be contiguous. However, when you justify the step, you should cite the step number of the conjuncts in the order they appear in the conjunction.

Here is a simple example of these two rules at work together. It is a proof of $C \wedge B \wedge C$ from $A \wedge B \wedge C$.

1. $A \land B \land C$	
2. B	$\land$ Elim: 1
3. C	$\land$ Elim: 1
4. $C \land B \land C$	$\land$ Intro: 3, 2, 3

This is a pretty silly thing to prove, but it does illustrate how our justifications are given. Notice that in step 4 we have to cite step 3 twice, since C appears twice in the conclusion.

In applying conjunction introduction, you will sometimes have to be careful about parentheses, due to our conventions about dropping parentheses. If one of the conjuncts is itself a conjunction, then of course there is no need to add any parentheses before forming the larger conjunction, unless you want to. For example, these are both correct applications of the rule:

Correct:

1. $P \land Q$	
2. R	
3. $(P \land Q) \land R$	$\land$ Intro: 1, 2

Correct:

1. $P \land Q$	
2. R	
3. $P \land Q \land R$	$\land$ Intro: 1, 2

However, if one of the conjuncts is a disjunction (or some other complex sentence), to prevent ambiguity, you may need to reintroduce parentheses that you omitted before. Thus the first of the following is a correct proof, but the second contains a faulty application of conjunction introduction, since it concludes with an ambiguous sentence.

Correct:

1. $P \lor Q$	
2. R	
3. $(P \lor Q) \land R$	$\land$ Intro: 1, 2

Wrong:

1. $P \lor Q$	
2. R	
3. $P \lor Q \land R$	$\land$ Intro: 1, 2

Disjunction rules

The rule of disjunction introduction allows you to go from a sentence P_i to any disjunction that has P_i among its disjuncts, say $P_1 \lor \ldots \lor P_i \lor \ldots \lor P_n$. In schematic form:

Disjunction Introduction (∨ Intro):

$$\begin{array}{l|l} & \mathsf{P}_i \\ & \vdots \\ \triangleright & \mathsf{P}_1 \vee \ldots \vee \mathsf{P}_i \vee \ldots \vee \mathsf{P}_n \end{array}$$

Once again, we stress that P_i may be the first or last disjunct of the conclusion. Further, as with conjunction introduction, some thought should be given about whether parentheses need be added to P_i to prevent ambiguity.

We now come to the first rule that corresponds to what we called a method of proof in the previous section. This is the rule of Disjunction Elimination or Proof by Cases. Recall that it allows you to conclude a sentence S from a disjunction $\mathsf{P}_1 \vee \ldots \vee \mathsf{P}_n$ if you can prove S from each of P_1 through P_n individually. The form of this rule requires us to discuss an important new structural feature of the Fitch-style system of deduction. This is the notion of a *subproof.* A subproof, as the name suggests, is a proof that occurs within the context of a larger proof. As with any proof, it may, and usually will, begin with an assumption, indicated as usual with a Fitch bar.

Before giving the schematic form of disjunction elimination, we give a particular formal proof that uses this rule. This will serve as a concrete illustration of how subproofs appear in $\mathcal{F}$.

```
| 1. (A ∧ B) ∨ (C ∧ D)
|---
|  | 2. A ∧ B
|  |---
|  | 3. B                      ∧ Elim: 2
|  | 4. B ∨ D                  ∨ Intro: 3
|
|  | 5. C ∧ D
|  |---
|  | 6. D                      ∧ Elim: 5
|  | 7. B ∨ D                  ∨ Intro: 6
| 8. B ∨ D                     ∨ Elim: 1, 2–4, 5–7
```

By making appropriate replacements for A, B, C, and D, this is a formalization of the proof given on page 62. It contains two subproofs. One of these shows that B ∨ D follows from A ∧ B, and runs from line 2 to 4. The other shows that the same conclusion follows from C ∧ D, and runs from line 5 to 7. These two proofs, together with the premise (A ∧ B) ∨ (C ∧ D) are just what we need to apply the rule of disjunction elimination. Hence these proofs become subproofs of our larger proof.

We now state the schematic version of the rule of disjunction elimination.

Disjunction Elimination ($\vee$ Elim):

```
| P1 ∨ ... ∨ Pn
| ⋮
| | P1
| |---
| | ⋮
| | S
| ⇓
| | Pn
| |---
| | ⋮
| | S
| ⋮
▷ | S
```

Let's look at another example of this rule, to emphasize how justifications involving subproofs are given. Let us prove that A follows from $(B \wedge A) \vee (A \wedge C)$.

```
| 1. (B ∧ A) ∨ (A ∧ C)
|---
| | 2. B ∧ A
| |---
| | 3. A                          ∧ Elim: 2
| | 4. A ∧ C
| |---
| | 5. A                          ∧ Elim: 4
| 6. A                            ∨ Elim: 1, 2–3, 4–5
```

In order to apply disjunction elimination, we need to cite a disjunctive sentence along with subproofs corresponding to each disjunct in that sentence. The citation for step 6 in our sample proof shows the form we use. The citation "n–m" is our way of referring to the subproof that begins on line n and ends on line m.

Sometimes, in using this rule, you will find it natural to use the Reiteration rule introduced in Chapter 2. For example, suppose we modify the above proof to show that A follows from $(B \wedge A) \vee A$.

1. $(B \land A) \lor A$	
2. $B \land A$	
3. A	$\land$ Elim: 2
4. A	
5. A	Reit: 4
6. A	$\lor$ Elim: 1, 2–3, 4–5

Here, the assumption of the second subproof is A, exactly the sentence we want to prove. So all we need to do is repeat that sentence to get the subproof into the desired form. (We could also just give a subproof with one step, but it is more natural to use Reiteration in such cases.)

Exercises and Problems

Problem 39. The following is an incomplete formal proof. It is missing some justifications and some sentences. These are indicated by question marks. Complete the proof by filling in the missing pieces.

1. $P \lor (Q \land R)$	
2. P	
3. $P \lor Q$	[?] Intro: 2
4. $P \lor R$	$\lor$ Intro: [?]
5. [?]	$\land$ Intro: 3, 4
6. $Q \land R$	
7. Q	$\land$ Elim: [?]
8. $P \lor Q$	$\lor$ Intro: 7
9. R	[?] Elim: 6
10. [?]	$\lor$ Intro: 9
11. $(P \lor Q) \land (P \lor R)$	$\land$ Intro: 8, 10
12. $(P \lor Q) \land (P \lor R)$	$\lor$ Elim: [?], 2–5, 6–[?]

Problem 40. Give formal proofs of the following:

1. $P \lor Q$ from the premise $P \land Q$
2. $a = c$ from the premise $(a = b) \land (b = c)$
3. $C \lor B$ from the premise $(A \land B) \lor C$
4. $(A \land B) \lor (A \land C)$ from the premises A and $B \lor C$

Negation rules

The rule of negation elimination corresponds to a very trivial valid step, from $\neg\neg P$ to P. Schematically:

Negation Elimination (¬ Elim):

```
|  ¬¬P
|   ⋮
▷|  P
```

The rule of negation introduction is, in contrast, one of the most interesting rules. It corresponds to the method of proof by contradiction. It too involves the use of a subproof, as will all our nontrivial methods of proof. The rule says that if you can prove a contradiction on the basis of an additional assumption P, then you are entitled to infer ¬P from the original premises. In formal systems it is traditional to require that the contradiction take the form of a proof of some sentence of the form Q ∧ ¬Q. Schematically:

Negation Introduction (¬ Intro):

```
| | P
| |---
| |  ⋮
| | Q ∧ ¬Q
▷| ¬P
```

Were we to loosen the requirement that our contradictions take this particular form, the system $\mathcal{F}$ would be considerably more natural. For example, we could get a contradiction by proving something like ¬(a = a) or Cube(c) ∧ Tet(c). In the exercises we will sometimes invite you to use this more general construal of a contradiction.

To illustrate this rule, let's show how to prove ¬¬A from A.

```
| 1. A
|---
| | 2. ¬A
| |---
| | 3. A ∧ ¬A                ∧ Intro: 1, 2
| 4. ¬¬A                     ¬ Intro: 2–3
```

Notice that in step 3, we have cited a step from outside the subproof, namely, step 1. This is legitimate, but raises an important issue. Just what steps *can* be cited at a given point in a proof? As a first guess, you might think that you can cite any earlier step. But this turns out to be wrong. We will explain why, and what the correct answer is, in the next section.

While we are on the topic of negation introduction, we use it to show how one can prove any sentence from contradictory premises.

1. P	
2. ¬P	
3. ¬Q	
4. P ∧ ¬P	∧ Intro: 1, 2
5. ¬¬Q	¬ Intro: 3–4
6. Q	¬ Elim: 5

Not all proofs begin with the assumption of premises. For example, if we wanted to prove some logical truth, we would start without any initial assumptions. Here is a proof of this sort, one that shows that ¬(P ∧ Q ∧ ¬P) is derivable in $\mathcal{F}$ without the use of premises.

1. P ∧ Q ∧ ¬P	
2. P	∧ Elim: 1
3. ¬P	∧ Elim: 1
4. P ∧ ¬P	∧ Intro: 2, 3
5. ¬(P ∧ Q ∧ ¬P)	¬ Intro: 1–4

Notice that there are no assumptions above the first horizontal Fitch bar, indicating that the main proof has no premises. The first step of the proof is the *subproof's* assumption. The subproof proceeds to derive a contradiction, based on this assumption, thus allowing us to conclude that the negation of the subproof's assumption follows without the need of premises. In other words, it is a logical truth.

The proper use of subproofs

Subproofs are the characteristic feature of Fitch-style deductive systems. It is important that you understand how to use them properly, since if you are not careful, you may "prove" things that don't follow from your premises. For example, the following formal proof looks like it is constructed according to our rules, but it purports to prove that A ∧ B follows from (B ∧ A) ∨ (A ∧ C), which is clearly not right.

1. (B ∧ A) ∨ (A ∧ C)	
2. B ∧ A	
3. B	∧ Elim: 2
4. A	∧ Elim: 2
5. A ∧ C	
6. A	∧ Elim: 5
7. A	∨ Elim: 1, 2–4, 5–6
8. A ∧ B	∧ Intro: 7, 3

The problem with this proof is step 8. In this step we have used step 3, a step that occurs within an earlier subproof. But it turns out that this sort of justification—one that reaches back inside a subproof that has already ended—is not legitimate. To understand why it's not legitimate, we need to think about what function subproofs play in a piece of reasoning.

A subproof typically looks something like this:

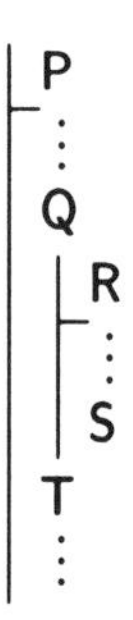

Subproofs begin with the introduction of a new assumption, in this example R. The reasoning within the subproof depends on this new assumption, together with any other premises or assumptions of the parent proof. So in our example, the derivation of S may depend on both P and R. When the subproof ends, indicated by the end of the vertical line that ties the subproof together, the subsequent reasoning can no longer use the subproof's assumption, or anything that depends on it. Nor can we use the Reiteration rule to get at those steps. We say that the assumption has been "discharged" or that the subproof has been "closed off."

When a subproof has been closed off, its individual steps are no longer accessible. It is only the subproof as a whole that can be cited as justification for some later step. What this means is that in justifying the assertion of T in our example, we could cite P, Q, and the subproof as a whole, but we could *not* cite individual items in the subproof like R or S. For these steps rely on assumptions we no longer have at our disposal. Once the subproof has been closed off, they are no longer accessible.

This, of course, is where we went wrong at step 8 in the fallacious proof given earlier. We cited a step in a subproof that had been closed off, namely, step 3. But the sentence at that step, B, had been proven on the basis of the assumption $B \wedge A$, an assumption we only made temporarily. The assumption is no longer in force at step 8, and so cannot be used at that point.

> **Remember**
>
> Once a subproof has been closed off, it can only be cited as a whole. Its individual items are not available.

This injunction does not prevent you from citing, from within a subproof, items that occur earlier outside the subproof, so long as they do not occur in subproofs that have been closed off. For example, in the schematic proof given above, the justification for S could well include the step that contains Q.

This observation becomes more pointed when you are working in a subproof of a subproof. We have not yet seen any examples where we needed to have subproofs within subproofs, but such examples are easy to come by. Here is one, which is a proof of one direction of the first DeMorgan law.

1. $\neg(P \wedge R)$	
2. $\neg(\neg P \vee \neg R)$	
3. $\neg P$	
4. $\neg P \vee \neg R$	$\vee$ Intro: 3
5. $(\neg P \vee \neg R) \wedge \neg(\neg P \vee \neg R)$	$\wedge$ Intro: 4, 2
6. $\neg\neg P$	$\neg$ Intro: 3–5
7. P	$\neg$ Elim: 6
8. $\neg R$	
9. $\neg P \vee \neg R$	$\vee$ Intro: 8
10. $(\neg P \vee \neg R) \wedge \neg(\neg P \vee \neg R)$	$\wedge$ Intro: 9, 2
11. $\neg\neg R$	$\neg$ Intro: 8–10
12. R	$\neg$ Elim: 11
13. $P \wedge R$	$\wedge$ Intro: 7, 12
14. $\neg(P \wedge R)$	Reit: 1
15. $(P \wedge R) \wedge \neg(P \wedge R)$	$\wedge$ Intro: 13, 14
16. $\neg\neg(\neg P \vee \neg R)$	$\neg$ Intro: 2–15
17. $\neg P \vee \neg R$	$\neg$ Elim: 16

Notice that the subproof 2–15 contains two subproofs, 3–5 and 8–10. In step 5 of the subproof 3–5 we cite step 2 from the parent subproof 2–14. Similarly, in step 10 of the subproof 8–10, we cite step 2. This is legitimate since the subproof 2–14 has not been closed off. While we did not need to in this proof, we could in fact have cited step 1 from the main proof in either of the sub-subproofs.

Another thing to note about this proof is the use of the Reiteration rule at step 14. We did not need to use Reiteration here, but did so just to illustrate a point. When it comes to subproofs, Reiteration is like any other rule: when you use it, you can cite steps outside of the immediate subproof, if the proofs that contain the cited steps have not yet been closed off. But you cannot cite a step in a closed subproof. For example, if we replaced the justification for step 10 with "Reit: 5," then our proof would no longer be correct.

Remember

In justifying a step of a subproof, you may cite any earlier step contained in the main proof, or in a subproof that has not yet been closed off.

A worked example

When giving formal proofs of the validity of arguments, you will discover that finding a formal proof is often far more difficult than recognizing the validity of the argument itself. An intuitively obvious inference may stump anyone not well versed in the tricks of the particular formal system. Different formal systems take different inferences as basic, and consequently what is easy in one may turn out to be a thorny problem in another.

No formal system can completely avoid these problems, and that is why formal proofs are never given in practice, even by professional logicians. Their main value lies not as tools for reasoning, but in theorizing about reasoning itself.

In this section, we will discuss how to go about finding a formal proof for a sample problem. The problem we will tackle is the other direction of the first DeMorgan law, showing that $\neg(\mathsf{P} \land \mathsf{Q})$ follows from $\neg\mathsf{P} \lor \neg\mathsf{Q}$.

A good way to approach the construction of a formal proof is to first convince yourself in detail that the conclusion really does follow from the premises. Then, see if you can mirror your reasoning within the formal system. Giving an informal proof is thus a good way to start. Here, then, is a detailed informal proof of the desired result.

> **Proof:** We are given that $\neg\mathsf{P} \lor \neg\mathsf{Q}$. For purposes of reductio, we will assume $\mathsf{P} \land \mathsf{Q}$ and attempt to derive a contradiction. There are two cases to consider, since we are given that either $\neg\mathsf{P}$ or $\neg\mathsf{Q}$ is true. But each of these contradicts the assumption $\mathsf{P} \land \mathsf{Q}$: $\neg\mathsf{P}$ contradicts the first conjunct and $\neg\mathsf{Q}$ contradicts the second. Consequently, our assumption leads to a contradiction, and so is false.

Let us now attempt to construct a formal proof which models this reasoning. As discussed in Chapter 2, we begin by writing our premises at the top of a skeleton proof, our desired conclusion at the bottom, leaving plenty of room in between.

```
| 1. ¬P ∨ ¬Q
|-
|  ⋮
| ¬(P ∧ Q)
```

The main method used in our informal proof was reductio, which corresponds to negation introduction. So we can fill in a bit more of the proof.

```
| 1. ¬P ∨ ¬Q
|-
| | 2. P ∧ Q
| |-
| |  ⋮
| | Some contradiction
| ¬(P ∧ Q)                            ¬ Intro: 2–?
```

Our informal proof showed that there was a contradiction whichever of ¬P or ¬Q was the case. The formal counterpart of proof by cases is disjunction elimination, and so our proof will look something like the following.

```
| 1. ¬P ∨ ¬Q
|-
| | 2. P ∧ Q
| |-
| | | 3. ¬P
| | |-
| | |  ⋮
| | | Some contradiction
| | | ¬Q
| | |-
| | |  ⋮
| | | Some contradiction
| | Some contradiction                ∨ Elim: 1, 3–?, ?–?
| ¬(P ∧ Q)                            ¬ Intro: 2–?
```

It is easy to find a contradiction in each of these cases.

Step	Justification
1. ¬P ∨ ¬Q	
2. P ∧ Q	
3. ¬P	
4. P	∧ Elim: 2
5. P ∧ ¬P	∧ Intro: 4, 3
6. ¬Q	
7. Q	∧ Elim: 2
8. Q ∧ ¬Q	∧ Intro: 7, 6
9. Some contradiction	∨ Elim: 1, 3–5, 6–8
¬(P ∧ Q)	¬ Intro: 2–?

If we try to finish this proof, we see that there is a problem at step 9. Disjunction elimination requires us to establish exactly the same sentence in both of the subproofs 3–5 and 6–8. Unfortunately, we have established two different contradictions. To give a proof in the system $\mathcal{F}$, we must somehow find a way to get around this problem, specifically, by deriving the same contradiction in each subproof. Let's try to establish P ∧ ¬P in each.

Step	Justification
1. ¬P ∨ ¬Q	
2. P ∧ Q	
3. ¬P	
4. P	∧ Elim: 2
5. P ∧ ¬P	∧ Intro: 4, 3
6. ¬Q	
7. Q	∧ Elim: 2
⋮	
P ∧ ¬P	
P ∧ ¬P	∨ Elim: 1, 3–5, 6–?
¬(P ∧ Q)	¬ Intro: 2–?

Here we need to use a trick. We have already shown, on page 72, how to derive anything from a contradiction in the system $\mathcal{F}$. Thus within our second subproof, we will use it's contradiction, Q ∧ ¬Q, to establish P ∧ ¬P. This will give us the same contradictory sentence in each subproof, and so will put us in a position to finish the proof.

1. $\neg P \vee \neg Q$	
2. $P \wedge Q$	
3. $\neg P$	
4. P	$\wedge$ Elim: 2
5. $P \wedge \neg P$	$\wedge$ Intro: 4, 3
6. $\neg Q$	
7. Q	$\wedge$ Elim: 2
8. $\neg(P \wedge \neg P)$	
9. $Q \wedge \neg Q$	$\wedge$ Intro: 7, 6
10. $\neg\neg(P \wedge \neg P)$	$\neg$ Intro: 8–9
11. $P \wedge \neg P$	$\neg$ Elim: 10
12. $P \wedge \neg P$	$\vee$ Elim: 1, 3–5, 6–11
13. $\neg(P \wedge Q)$	$\neg$ Intro: 2–12

In the next chapter, we will introduce a streamlined version of $\mathcal{F}$ that will allow our formal proofs to more closely mirror informal reasoning. We will illustrate it by reproving this result.

Exercises and Problems

Problem 41. The following is an incomplete formal proof. It is missing some justifications and some sentences. These are indicated by question marks. Complete the proof by filling in the missing pieces.

1. $P \vee Q$	
2. $\neg P$	
3. [?]	
4. $\neg Q$	
5. $P \wedge \neg P$	$\wedge$ Intro: [?]
6. $\neg\neg Q$	$\neg$ Intro: 4–5
7. [?]	$\neg$ Elim: 6
8. Q	
9. Q	Reit: 8
10. Q	$\vee$ Elim: 1, [?], 8–9

Problem 42. The following is an incomplete proof of $P \vee \neg P$, showing that we can prove this without any premises. It is missing some justifications and some sentences. These are indicated by question marks. Complete the proof by filling in the missing pieces.

1. $\neg(P \lor \neg P)$	
2. P	
3. ?	$\lor$ Intro: 2
4. $(P \lor \neg P) \land \neg(P \lor \neg P)$	$\land$ Intro: 3, 1
5. $\neg P$	$\neg$ Intro: 2–4
6. ?	
7. $P \lor \neg P$	$\lor$ Intro: 6
8. $(P \lor \neg P) \land \neg(P \lor \neg P)$	?
9. $\neg\neg P$	$\neg$ Intro: 6–8
10. $\neg P \land \neg\neg P$	$\land$ Intro: 5, 9
11. $\neg\neg(P \lor \neg P)$	$\neg$ Intro: 1–10
12. $P \lor \neg P$	?

Problem 43. Prove $P \lor (Q \land R)$ from the premises $P \lor Q$ and $P \lor R$. This will require you to use a subproof within a subproof.

Problem 44. Give an informal proof that might have been used by the authors in constructing the formal proof shown on page 75.

Problem 45. Give a formal proof mirroring the informal proof on page 63 of $\neg(b = c)$ from the premises Cube(c) $\lor$ Dodec(c) and Tet(b). You will have to use the more general notion of contradiction discussed on page 72.

Problem 46. Give informal proofs of the following (without presupposing the DeMorgan laws). Then give formal proofs that mirror your informal proofs as closely as possible.

1. $\neg P \land \neg Q$ from the premise $\neg(P \lor Q)$
2. $\neg(P \lor Q)$ from the premise $\neg P \land \neg Q$

Problem 47. Give formal proofs of the following:

1. $\neg A$ from the premise $\neg(A \lor B)$
2. A from the premises $\neg(\neg A \land B)$ and $\neg(\neg B \lor C)$
3. $\neg(a = b \land b \neq a)$ from no premises

Citing theorems

Proofs in the system $\mathcal{F}$ can get extremely long, especially if we insist on going back to first principles in every proof. In your high school geometry class, you were allowed to cite previously established theorems in giving subsequent proofs, and this shortened the later proofs considerably.

We allow a similar use of earlier theorems in system $\mathcal{F}$. If you have already constructed a proof of some conclusion Q from premises

$P_1, \ldots, P_n$, then you should certainly be allowed to assert Q when you have established $P_1, \ldots, P_n$. When we do this, we will cite as justification the steps where $P_1, \ldots, P_n$ appear in the proof, and indicate that this is a previous theorem.

Here is an example of a very short formal proof which exploits three previous proofs.

1.	¬(P ∧ Q)	
2.	P	
3.	¬P ∨ ¬Q	Prev Thm (page 75): 1
4.	¬¬P	Prev Thm (page 72): 2
5.	¬Q	Prev Thm (Problem 41): 3, 4

If you look back to the proofs cited here, you will notice that there is a mismatch in the sentence letters used. For example on page 72, we showed that ¬¬A followed from A, but in our proof we are concluding ¬¬P from P. This is permissible, since the earlier proof would be just as correct if A were replaced by P, or indeed by any other sentence, simple or complex.

An important special case of citing previous proofs arises when you cite a proof that involves no premises. For example in Problem 42, we showed that P ∨ ¬P can be proved without premises. In the following proof, we cite this previous theorem.

1.	(P ∧ Q) ∨ ¬(P ∧ Q)	Prev Thm (Problem 42)
2.	P ∧ Q	
3.	(P ∧ Q) ∨ ¬P ∨ ¬Q	∨ Intro: 2
4.	¬(P ∧ Q)	
5.	¬P ∨ ¬Q	Prev Thm (page 75): 4
6.	(P ∧ Q) ∨ ¬P ∨ ¬Q	∨ Intro: 5
7.	(P ∧ Q) ∨ ¬P ∨ ¬Q	∨ Elim: 1, 2–3, 4–6

Notice that here, we have replaced the P in our previous theorem by a complex formula (P ∧ Q).

Exercises and Problems

Problem 48. In order to appreciate the utility of citing earlier proofs, write out a complete proof of (P ∧ Q) ∨ ¬P ∨ ¬Q, one that does not cite any previous theorem. Compare this proof to the one given immediately above.

Problem 49. Give a formal proof of Happy(Carl) from the following three premises:

1. ¬(Home(Max) ∧ Home(Claire))
2. Home(Max) ∨ Happy(Carl)
3. Home(Claire) ∨ Happy(Carl)

Compare the resulting formal proof with the informal proof you gave in Problem 29. The results should be fairly similar. However, if you were not allowed to cite earlier results in your formal proof, it would be quite long and tedious.

Problem 50. Give a formal proof of Small(c) from the two premises Cube(c) ∨ Small(c) and Dodec(c). You will need to use the more general construal of contradiction discussed on page 72.

3.10 Conjunctive and disjunctive normal forms

We learned earlier how to start with a sentence and transform it into a logically equivalent sentence in negation normal form, one where all negations occur in front of atomic sentences. We can improve on this by introducing the so-called distributive laws. These are equivalences that will allow us to transform sentences into what are known as "conjunctive normal form" and "disjunctive normal form." These normal forms are quite important in applications of logic in computer science, as we discuss in Chapter 10.

Recall that in algebra you learned that multiplication distributes over addition: $a \times (b + c) = (a \times b) + (a \times c)$. The distributive laws of logic look formally much the same. One version tells us that P ∧ (Q ∨ R) is logically equivalent to (P ∧ Q) ∨ (P ∧ R). That is, ∧ distributes over ∨. To see that this is so, notice that the first sentence is true if and only if P plus at least one of Q or R is true. But a moment's thought shows that the second sentence is true in exactly the same circumstances.

In arithmetic, + does not distribute over ×. However, ∨ does distribute over ∧. That is to say, P ∨ (Q ∧ R) is logically equivalent to (P ∨ Q) ∧ (P ∨ R).

Remember

(The distributive laws) For any sentences P, Q, and R:

1. Distribution of ∧ over ∨: P ∧ (Q ∨ R) ⇔ (P ∧ Q) ∨ (P ∧ R)
2. Distribution of ∨ over ∧: P ∨ (Q ∧ R) ⇔ (P ∨ Q) ∧ (P ∨ R)

As you recall from algebra, the distributive law for × over + is incredibly useful. It allows us to transform any algebraic expression involving + and ×, no matter how complex, into one that is just a sum of products. For example, the following transformation uses distribution three times.

$$\begin{aligned}(a+b)(c+d) &= (a+b)c+(a+b)d \\ &= ac+bc+(a+b)d \\ &= ac+bc+ad+bd\end{aligned}$$

In exactly the same way, the distribution of $\land$ over $\lor$ allows us to transform any sentence built up from literals by means of $\land$ and $\lor$ into a logically equivalent sentence that is a disjunction of (one or more) conjunctions of (one or more) literals. That is, using this first distributive law, we can turn any sentence in negation normal form into a sentence in negation normal form that is a disjunction of conjunctions of literals. A sentence in this form is said to be in *disjunctive normal form.*

Here is an example that parallels our algebraic example. Notice that, as in the algebraic example, we are distributing in from the right as well as the left, even though our statement of the rule was only on the left.

$$\begin{aligned}(\mathsf{A} \lor \mathsf{B}) \land (\mathsf{C} \lor \mathsf{D}) &\Leftrightarrow [(\mathsf{A} \lor \mathsf{B}) \land \mathsf{C}] \lor [(\mathsf{A} \lor \mathsf{B}) \land \mathsf{D}] \\ &\Leftrightarrow (\mathsf{A} \land \mathsf{C}) \lor (\mathsf{B} \land \mathsf{C}) \lor [(\mathsf{A} \lor \mathsf{B}) \land \mathsf{D}] \\ &\Leftrightarrow (\mathsf{A} \land \mathsf{C}) \lor (\mathsf{B} \land \mathsf{C}) \lor (\mathsf{A} \land \mathsf{D}) \lor (\mathsf{B} \land \mathsf{D})\end{aligned}$$

Likewise, using distribution of $\lor$ over $\land$, we can turn any negation normal form sentence into one that is a *conjunction* of one or more sentences, each of which is a *disjunction* of one or more literals. A sentence in this form is said to be in *conjunctive normal form.*

The parallel example to that given above has $\land$ and $\lor$ interchanged:

$$\begin{aligned}(\mathsf{A} \land \mathsf{B}) \lor (\mathsf{C} \land \mathsf{D}) &\Leftrightarrow [(\mathsf{A} \land \mathsf{B}) \lor \mathsf{C}] \land [(\mathsf{A} \land \mathsf{B}) \lor \mathsf{D}] \\ &\Leftrightarrow (\mathsf{A} \lor \mathsf{C}) \land (\mathsf{B} \lor \mathsf{C}) \land [(\mathsf{A} \land \mathsf{B}) \lor \mathsf{D}] \\ &\Leftrightarrow (\mathsf{A} \lor \mathsf{C}) \land (\mathsf{B} \lor \mathsf{C}) \land (\mathsf{A} \lor \mathsf{D}) \land (\mathsf{B} \lor \mathsf{D})\end{aligned}$$

On page 45 we showed how to transform the sentence $\neg((\mathsf{A} \lor \mathsf{B}) \land \neg\mathsf{C})$ into one in negation normal form. The result was $(\neg\mathsf{A} \land \neg\mathsf{B}) \lor \mathsf{C}$. This sentence just happens to be in disjunctive normal form. Let us repeat our earlier transformation, but continue until we get a sentence in conjunctive normal form.

$$\begin{aligned}\neg((\mathsf{A} \lor \mathsf{B}) \land \neg\mathsf{C}) &\Leftrightarrow \neg(\mathsf{A} \lor \mathsf{B}) \lor \neg\neg\mathsf{C} \\ &\Leftrightarrow (\neg\mathsf{A} \land \neg\mathsf{B}) \lor \neg\neg\mathsf{C} \\ &\Leftrightarrow (\neg\mathsf{A} \land \neg\mathsf{B}) \lor \mathsf{C} \\ &\Leftrightarrow (\neg\mathsf{A} \lor \mathsf{C}) \land (\neg\mathsf{B} \lor \mathsf{C})\end{aligned}$$

It is important to remember that a sentence can count as being in both conjunctive and disjunctive normal forms at the same time. For example, the sentence Home(Claire) $\land$ $\neg$Home(Max) is such a sentence. On the one hand it is in disjunctive normal form since it is a disjunction of one sentence (itself) which is a conjunction of two literals. On the other hand it is in conjunctive normal form since it is a conjunction of two sentences, each of which is a disjunction of one literal.

Remember

1. A sentence is in disjunctive normal form (DNF for short) if it is a disjunction of one or more conjunctions of one or more literals.
2. A sentence is in conjunctive normal form (CNF for short) if it is a conjunction of one or more disjunctions of one or more literals.
3. Distribution of $\wedge$ over $\vee$ allows you to transform any sentence in negation normal form into one in DNF.
4. Distribution of $\vee$ over $\wedge$ allows you to transform any sentence in negation normal form into one in CNF.
5. Some sentence are in both CNF and DNF.

Exercises and Problems

Exercise$_\diamond$ 51. (Distributive laws) Open Distributive Sentences. Each even numbered sentence has been obtained from the sentence above it by applying a distributive law. Build any world in which all the names are used, and check to see that each odd numbered sentence has the same truth value in the world as the even numbered sentence which follows it.

Problem$_\diamond$ 52. Open CNF Sentences. In this file you will find some conjunctive normal form sentences in the odd numbered positions, but you will see that the even numbered sentences are blank. You should fill in a DNF sentence logically equivalent to the sentence above it. Save the file as Sentences 3.52. Check your work by opening several worlds and checking to see that each of your sentences has the same truth value as the one from which you obtained it.

Problem$_\diamond$ 53. Open More CNF Sentences. In this file you will find sentences in every third position. The two blanks are for you to first transform the sentence into negation normal form, and then put that sentence into CNF. Save the file as Sentences 3.53. Again, check your work by opening several worlds to see that each of your sentences has the same truth value as the original.

Problem 54. Assume that A, B, C, and D are literals. Write each of the following in disjunctive normal form. That is, use distribution of $\wedge$ over $\vee$ to find a logically equivalent expression in DNF. Simplify your answer as much as possible using the laws of commutativity and idempotence of $\wedge$ and $\vee$.

1. C ∧ (A ∨ (B ∧ C))
2. B ∧ (A ∧ B ∧ (A ∨ B ∨ (B ∧ C)))
3. A ∧ (A ∧ (B ∨ (A ∧ C)))

Problem 55. Give a formal proof in the system $\mathcal{F}$ of the following:

1. P ∧ (Q ∨ R) from the premise (P ∧ Q) ∨ (P ∧ R).
2. (P ∧ Q) ∨ (P ∧ R) from the premise P ∧ (Q ∨ R).
3. P ∨ (Q ∧ R) from the premise (P ∨ Q) ∧ (P ∨ R).
4. (P ∨ Q) ∧ (P ∨ R) from the premise P ∨ (Q ∧ R).

Problem 56. (More practice on normal forms) For each of the following sentences, first find a sentence in negation normal form that is logically equivalent to it. Then take the result and transform it into both CNF and DNF. Assume that A, B, C, and D are atomic sentences.

1. ¬[A ∧ (B ∨ ¬C)]
2. A ∨ ¬(C ∨ B)
3. ¬[(¬A ∧ ¬B) ∨ (¬C ∧ D)]
4. A ∧ ¬(C ∨ D) ∧ B

Problem$_\Diamond$ 57. In general, it is much easier to tell at a glance whether a sentence in CNF or DNF is true than it is for sentences not in normal form. This is reflected in the game, which lasts at most three moves for a sentence in one of these normal forms, one move for each of ∧, ∨, and ¬. Which of these normal forms is the most perspicuous in a given instance depends on the sentence. Sometimes one can be much longer than the other. Open CNF/DNF Sentences. Put each odd numbered sentence into CNF or DNF in the space below it, depending on its form. (That is, put the ones in CNF into DNF and vice versa.) Then build a world in which all of the original sentences are true. Verify that all the new sentences are also true. Save the world as World 3.57 and the sentences as Sentences 3.57.

3.11 Truth-functional completeness

So far we have introduced into our language three truth-functional connectives, ¬, ∧, and ∨. We could introduce many others besides these, of course. For example, we could introduce a symbol for *neither... nor* or to express exclusive disjunction. We have seen, however, that these notions can already be expressed using the three connectives at our disposal. This makes one wonder whether there are any binary truth-functional connectives that can't be expressed with these three connectives.

How many binary truth-functional connectives are imaginable? If you think about the possible truth tables for binary connectives, you can compute the number of binary connectives. First, since we are dealing with binary connectives, there are four rows. Each row can be assigned either TRUE or FALSE, so there are $2^4 = 16$ ways of doing this. We could look at each of these truth tables in turn and show how to capture it with the existing connectives, just as we captured *neither A nor B* with $\neg(A \vee B)$. But there is a simpler way to show this.

Any binary truth-functional connective, say $\star$, that we might introduce would have a truth table like the following, with one of the values TRUE or FALSE in each row.

P	Q	P ⋆ Q
TRUE	TRUE	*1st value*
TRUE	FALSE	*2nd value*
FALSE	TRUE	*3rd value*
FALSE	FALSE	*4th value*

Let us consider how we might express this truth function, no matter what the four values happen to be.

First, if all four values are FALSE, then we can clearly express $P \star Q$ with the sentence $P \wedge \neg P \wedge Q \wedge \neg Q$. But suppose at least one of the values is TRUE. How can we express $P \star Q$? One way would be this. Let us use $C_1, \ldots, C_4$ for the following four conjunctions:

$$\begin{aligned} C_1 &= (P \wedge Q) \\ C_2 &= (P \wedge \neg Q) \\ C_3 &= (\neg P \wedge Q) \\ C_4 &= (\neg P \wedge \neg Q) \end{aligned}$$

Notice that sentence C_1 will be TRUE if the truth values of P and Q are as specified in the first row of the truth table, and that if the values of P and Q are anything else, then C_1 will be false. Similarly with C_2 and the second row of the truth table, and so forth. To get a sentence that gets the value TRUE in exactly the same rows as $P \star Q$, all we need do is take the disjunction of the appropriate C's. For example, if $P \star Q$ is true in rows 2 and 4, then $C_2 \vee C_4$ is equivalent to this sentence.

What this shows is that all possible binary truth functional connectives are expressible using just the connectives $\neg$, $\wedge$, and $\vee$. In fact, it shows that they can all be expressed using sentences in DNF.

It is easy to see that a similar procedure allows us to express all possible unary truth functional connectives. Here, a unary connective, say $\natural$, will have a truth table like this:

P	♮P
TRUE	*1st value*
FALSE	*2nd value*

If both of the values under $\natural P$ are FALSE, then we can express it using the sentence $P \wedge \neg P$. Otherwise, we can express $\natural P$ as a disjunction of one or more of the following:

$$\begin{aligned} \mathsf{C}_1 &= \mathsf{P} \\ \mathsf{C}_2 &= \neg\mathsf{P} \end{aligned}$$

C_1 will be one of the disjuncts if the first value is TRUE, and C_2 will be if the second value is TRUE.

Once we see how this procedure is working, we see that it would apply equally well to truth functional connectives of any arity. Suppose, for example, that we had a truth functional connective whose truth table looked like this:

P	Q	R	If P then Q, else R
TRUE	TRUE	TRUE	TRUE
TRUE	TRUE	FALSE	TRUE
TRUE	FALSE	TRUE	FALSE
TRUE	FALSE	FALSE	FALSE
FALSE	TRUE	TRUE	TRUE
FALSE	TRUE	FALSE	FALSE
FALSE	FALSE	TRUE	TRUE
FALSE	FALSE	FALSE	FALSE

A fairly good English translation of this is *if P then Q, else R*. When we apply the above method to express this connective, we get the following sentence:

$$(\mathsf{P} \land \mathsf{Q} \land \mathsf{R}) \lor (\mathsf{P} \land \mathsf{Q} \land \neg\mathsf{R}) \lor (\neg\mathsf{P} \land \mathsf{Q} \land \mathsf{R}) \lor (\neg\mathsf{P} \land \neg\mathsf{Q} \land \mathsf{R})$$

We will say that the operators $\neg$, $\land$, and $\lor$ are *truth functionally complete*, since they allow us to express any conceivable truth functional connective. There are other collections of operators that are truth functionally complete. In fact, we could get rid of either $\land$ or $\lor$ without losing truth functional completeness. For example, $\mathsf{P} \lor \mathsf{Q}$ can be expressed using just $\neg$ and $\land$ as follows:

$$\neg(\neg\mathsf{P} \land \neg\mathsf{Q})$$

What this means is that we could get rid of all the occurrences of $\lor$ in our sentences in favor of $\neg$ and $\land$. Of course the result would be much longer and harder to understand.

In the next chapter, we will introduce two additional binary connectives, $\rightarrow$ and $\leftrightarrow$. While we now know that these connectives do not result in any increase in expressive power, having them will often simplify our sentences considerably.

Exercises and Problems

Problem 58. (Unary connectives) For each of the columns in the following truth table, find the sentence that the above method would give you.

P				
TRUE	TRUE	TRUE	FALSE	FALSE
FALSE	TRUE	FALSE	TRUE	FALSE

Problem 59. (Expressing another ternary connective) Use the method we have developed to express the ternary connective ♡ defined in the following truth table. Then see if you can simplify the result as much as possible.

P	Q	R	♡(P, Q, R)
TRUE	TRUE	TRUE	TRUE
TRUE	TRUE	FALSE	TRUE
TRUE	FALSE	TRUE	TRUE
TRUE	FALSE	FALSE	FALSE
FALSE	TRUE	TRUE	FALSE
FALSE	TRUE	FALSE	TRUE
FALSE	FALSE	TRUE	TRUE
FALSE	FALSE	FALSE	TRUE

Problem* 60. (Simplifying *if... then... else*) Assume that P, Q, and R are atomic sentences. See if you can simplify the sentence we came up with to express if P then Q, else R, so that it becomes a disjunction of two sentences, each of which is a conjunction of two literals.

Problem* 61. Suppose we had introduced $\underline{\vee}$ for exclusive disjunction. State valid introduction and elimination rules for this connective. [Hint: What are the informal ways of proving or using such statements in proofs?]

3.12 Alternative notation

As we mentioned in the last chapter, there are various dialect differences among users of FOL. It is important to be aware of these so that you will not be stymied by superficial differences. In fact, you will run into alternate symbols being used for each of the three connectives studied in this chapter.

The most common variant of the negation sign, ¬, is the symbol known as the tilde, ∼. Thus you will frequently encounter ∼P where we would write ¬P. A more old-fashioned alternative is to draw a bar completely across the negated sentence, as in $\overline{\mathsf{P}}$. This has one advantage over ¬, in that it allows you to avoid certain uses of parentheses, since the bar indicates its own scope by what lies under it. For example, where we have to write ¬(P ∧ Q), the bar equivalent would simply be $\overline{\mathsf{P} \land \mathsf{Q}}$. None of these symbols are available on all keyboards, a serious problem in some contexts, such as programming languages. Because of this, many programming languages use an exclamation point to indicate negation. In the language C, for example, $\neg P$ would be written !P.

There are only two common variants of $\wedge$. By far the most common is &, or sometimes (as in C), &&. An older notation uses a centered dot, as in multiplication. To make things more confusing still, the dot is sometimes omitted, again as in multiplication. Thus, for $P \wedge Q$ you might see any of the following: P&Q, P&&Q, $P \cdot Q$, or just PQ.

Happily, the symbol $\vee$ is pretty standard. The only exception you may encounter is a single or double vertical line, used in programming languages. So if you see $P \mid Q$ or $P \parallel Q$, what is meant is probably $P \vee Q$. Unfortunately, though, some textbooks use $P \mid Q$ to express *not both P and Q*.

Alternatives to parentheses

There are ways to get around the use of parentheses in FOL. At one time, a common alternative to parentheses was a system known as dot notation. This system involved placing little dots next to connectives indicating their relative "power" or scope. In this system, the two sentences we write as $P \vee (Q \wedge R)$ and $(P \vee Q) \wedge R$ would have been written $P \vee .\ Q \wedge R$ and $P \vee Q\ . \wedge R$, respectively. With more complex sentences, multiple dots were used. Fortunately, this notation has just about died out, and the present authors never speak to anyone who uses it.

Another approach to parentheses is known as Polish notation. In Polish notation, the usual infix notation is replaced by prefix notation, and this makes parentheses unnecessary. Thus the distinction between our $\neg(P \vee Q)$ and $(\neg P \vee Q)$ would, in prefix form, come out as $\neg \vee PQ$ and $\vee \neg PQ$, the order of the connectives indicating which includes the other in its scope.

Besides prefix notation, Polish notation uses certain capital letters for connectives (N for $\neg$, K for $\wedge$, and A for $\vee$), and lower case letters for its atomic sentences (to distinguish them from connectives). So an actual sentence of the Polish dialect would look like this:

$$ApKNqr$$

Since this expression starts with A, we know right away that it is a disjunction. What follows must be its two disjuncts, in sequence. So the first disjunct is p and the second is KNqr, that is, the conjunction of the negation of q and of r. So this is the Polish version of

$$P \vee (\neg Q \wedge R)$$

Though Polish notation may look hard to read, many of you have already mastered a version of it. Calculators use two styles for entering formulas. One is known as algebraic style, the other as RPN style. The

RPN stands for "reverse Polish notation." If you have a calculator that uses it, then to calculate the value of, say, $(7 \times 8) + 3$ you enter things in this order: 7, 8, ×, 3, +. This is just the reverse of the Polish, or prefix, ordering.

In order for Polish notation to work without parentheses, the connectives must all have a fixed arity. If we allowed conjunction to take an arbitrary number of sentences as arguments, rather than requiring exactly two, a sentence like KpNKqrs would be ambiguous. It could either mean $P \wedge \neg(Q \wedge R) \wedge S$ or $P \wedge \neg(Q \wedge R \wedge S)$, and these aren't equivalent.

Remember

The following table summarizes the alternative notations discussed so far.

Our notation	Common equivalents
$\neg P$	$\sim P$, $\overline{P}$, !P, Np
$P \wedge Q$	P&Q, P&&Q, $P \cdot Q$, PQ, Kpq
$P \vee Q$	P | Q, P || Q, Apq

Exercises and Problems

Problem 62. (Overcoming dialect differences) The following are all sentences of FOL. But they're in different dialects. Translate them into our dialect.

1. $\overline{\overline{P}\&\overline{Q}}$
2. !(P || (Q&&P))
3. $(\sim P \vee Q) \cdot P$
4. $P(\sim Q \vee RS)$

Problem 63. (Translating from Polish) Try your hand at translating the following sentences from Polish notation into our dialect.

1. NKpq
2. KNpq
3. NAKpqArs
4. NAKpAqrs
5. NAKApqrs

4

Conditionals and Biconditionals

In this chapter we introduce two additional truth functional connectives, the material conditional and the material biconditional. In the next chapter we will see that the first of these is of great use in expressing certain kinds of "quantification." The second, the biconditional, is of secondary importance.

4.1 Material conditional symbol ($\rightarrow$)

This symbol is used to combine two sentences P and Q to form a new sentence $P \rightarrow Q$, called a *material conditional.* The sentence P is called the *antecedent* of the conditional, and Q is called the *consequent* of the conditional. We will discuss the English counterparts of this symbol after we explain its meaning.

Semantics and the game rule for the conditional

The sentence $P \rightarrow Q$ is true if and only if either P is false or Q is true (or both). This can be summarized by the following truth table.

P	Q	$P \rightarrow Q$
TRUE	TRUE	TRUE
TRUE	FALSE	FALSE
FALSE	TRUE	TRUE
FALSE	FALSE	TRUE

A second's thought shows that $P \rightarrow Q$ is really just another way of saying $\neg P \vee Q$. Tarski's World in fact treats the former as an abbreviation of the latter. In particular, in playing the game, Tarski's World simply replaces a statement of the form $P \rightarrow Q$ by its equivalent $\neg P \vee Q$.

Remember

The best way to think of $P \rightarrow Q$ is to remember that this conditional statement is false only in one case: when the antecedent P is true but the consequent Q is false.

English forms of the material conditional

We can come fairly close to an adequate English rendering of the conditional expression $\mathsf{P} \rightarrow \mathsf{Q}$ with the sentence *If P then Q*. At any rate, it is clear that this English conditional, like the material conditional, is false if *P* is true and *Q* is false. Thus, we will translate, for example, *If Max is home then Claire is at the library* as:

$$\mathsf{Home(Max)} \rightarrow \mathsf{Library(Claire)}$$

Other English expressions that can frequently be translated using the material conditional $\mathsf{P} \rightarrow \mathsf{Q}$ include: *P only if Q, Q provided P,* and *Q when P*. We also use $\rightarrow$ in combination with $\neg$ to translate sentences of the form *Unless P, Q* or *Q unless P* . These mean the same thing as *Q if not P*, and so are translated as $\neg\mathsf{P} \rightarrow \mathsf{Q}$.

In this course we will always translate the English *if... then* using $\rightarrow$, but there are in fact many uses of the English expression that cannot be adequately expressed using the material conditional. For example, the sentence,

If Max had been at home, then Carl would have been there too,

can be false even if Max was not in fact at home. But the first-order sentence,

$$\mathsf{Home(Max)} \rightarrow \mathsf{Home(Carl)}$$

is automatically true if Max is not at home. The point of such examples is that many uses of *if... then* just aren't truth functional. The truth of the whole depends on something more than the truth values of the parts. It seems to depend on there being some genuine connection between the subject-matter of the antecedent and that of the consequent. But these are controversial matters. We will take them up a bit further in Section 4.3.

The most important use of $\rightarrow$ in first-order logic is not in conjunction with the above expressions, but rather with universally quantified sentences, sentences of the form *All A's are B's* and *Every A is a B*. The analogous first-order sentences have the form:

$$\text{For every object x } (\mathsf{A(x)} \rightarrow \mathsf{B(x)})$$

This says that any object you pick will either fail to be an A or else be a B. We will discuss such sentences in more detail later, once we have variables and the symbol $\forall$ at our disposal.

Logical truth and logical consequence

There is one other thing we should say about the material conditional, which helps explain its importance in logic. The conditional allows us to reduce the notion of logical consequence to that of logical truth, at least in cases where we have only finitely many premises. We said that a sentence Q is a consequence of premises $P_1, \ldots, P_n$ if and only if it is impossible for all the premises to be true while the conclusion is false. Another way of saying this is that it is impossible for the single sentence $(P_1 \wedge \ldots \wedge P_n)$ to be true while Q is false.

Given the meaning of $\rightarrow$, we see that Q is a consequence of $P_1, \ldots, P_n$ if and only if it is impossible for the single sentence

$$(P_1 \wedge \ldots \wedge P_n) \rightarrow Q$$

to be false, that is, just in case this conditional sentence is logically true. Thus, one way to check for the validity of an argument in propositional logic, at least in theory, is to construct a truth table for this sentence and see whether the final column contains only TRUE. In practice, this method is not very practical, since the truth tables quickly get too large to be manageable.

4.2 Biconditional symbol ($\leftrightarrow$)

Our final connective is the material biconditional symbol. Given any sentences P and Q there is another sentence formed by connecting these by means of the biconditional: $P \leftrightarrow Q$. A sentence of the form $P \leftrightarrow Q$ is true if and only if P and Q have the same truth value, that is, either they are both true or both false. In English this is commonly expressed using the expression *if and only if*,[1] and, in mathematical discourse, *just in case*. So, for example, the sentence *Max is home if and only if Carl is home* would be translated as:

$$\text{Home(Max)} \leftrightarrow \text{Home(Carl)}$$

The mathematical sentence *n is even just in case* n^2 *is even* would likewise be translated as:

$$Even(n) \leftrightarrow Even(n^2)$$

An important fact about the biconditional symbol is that two sentences P and Q are logically equivalent if and only if the biconditional formed from them, $P \leftrightarrow Q$, is logically true. Another way of putting this is to say that $P \Leftrightarrow Q$ is true if and only if the FOL sentence $P \leftrightarrow Q$ is *logically* true. So, for example, we can express one of the DeMorgan laws by saying that the following sentence is logically true:

$$\neg(P \vee Q) \leftrightarrow (\neg P \wedge \neg Q)$$

This observation makes it tempting to confuse the symbols $\leftrightarrow$ and $\Leftrightarrow$. This temptation must be resisted. The former is a truth-functional connective of FOL, while the latter is an abbreviation of "is logically equivalent to."

[1]Mathematicians often write "iff" for "if and only if."

Semantics and the game rule for $\leftrightarrow$

The semantics for the biconditional can be summarized with the following truth table.

P	Q	P $\leftrightarrow$ Q
TRUE	TRUE	TRUE
TRUE	FALSE	FALSE
FALSE	TRUE	FALSE
FALSE	FALSE	TRUE

Notice that the final column of this truth table is the same as that for $(P \rightarrow Q) \wedge (Q \rightarrow P)$. (See Problem 1 below.) For this reason, logicians often treat a sentence of the form $P \leftrightarrow Q$ as an abbreviation of $(P \rightarrow Q) \wedge (Q \rightarrow P)$. Tarski's World also uses this abbreviation in the game. Thus, the game rule for $P \leftrightarrow Q$ is simple. Whenever a sentence of this form is encountered, it is replaced by $(P \rightarrow Q) \wedge (Q \rightarrow P)$.

Exercises and Problems

Problem 1. Construct a truth table for the sentence $(A \rightarrow B) \wedge (B \rightarrow A)$. Show that its final column is the same as that for $A \leftrightarrow B$.

Exercise$_\diamond$ 2. (Evaluating sentences in a world) Run through Abelard's Sentences, evaluating them in Wittgenstein's World. If you make a mistake, play the game to see where you have gone wrong.

Problem$_\diamond$ 3. (Describing a world) Choose **Hide labels** from the **Display** menu, and then open Montague's World. In this world, each object has a name, and no object has more than one name. Start a new sentence file where you will describe some features of this world. Check each of your sentences to see that it is indeed a sentence and that it is true in this world.

1. Notice that if c is a tetrahedron, then a is not a tetrahedron. (Remember, in this world each object has exactly one name.) Use your first sentence to express this fact.
2. However, note that the same is true of b and d. That is, if b is a tetrahedron, then d isn't. Use your second sentence to express this.
3. Finally, observe that if b is a tetrahedron, then c isn't. Express this.
4. Notice that if a is a cube and b is a dodecahedron, then a is to the left of b. Use your next sentence to express this fact.
5. Use your fifth sentence to express the fact that d is a tetrahedron if and only if it is small.
6. Next, express the fact that if a and d are both cubes, then one is

to the left of the other. [Note: You will need to use a disjunction to express the fact that one is to the left of the other.]

7. Notice that d is a cube if and only if it is either medium or large. Express this.
8. Observe that if b is neither to the right nor left of d, then one of them is a tetrahedron. Express this observation.
9. (⋆) Finally, express the fact that b and c are the same size if and only if one is a tetrahedron and the other is a dodecahedron.

Save your sentences as Sentences 4.3 Now choose **Show labels** from the **Display** menu. Verify that all of your sentences are indeed true. When verifying the first three, pay particular attention to the truth values of the various constituents. Notice that sometimes the conditional has a false antecedent and sometimes a true consequent. What it never has is a true antecedent and a false consequent. In each of these three cases, play the game committed to true. Make sure you understand why the game proceeds as it does.

Problem$_\diamond$ **4.** (Translation) Translate the following English sentences into FOL. Your translations will use all of the propositional connectives.

1. *If a is a tetrahedron then it is in front of d.*
2. *a is to one side or the other of d only if it's a cube.*
3. *c is between either a and e or a and d.*
4. *c is to the right of a if it (i.e., c) is small.*
5. *c is to the right of d only if b is to the right of c and left of e.*
6. *If e is a tetrahedron, then it's to the right of b if and only if it is also in front of b.*
7. *If b isn't in front of d then neither is it in back of d, provided it's a cube.*
8. *c is in back of a but in front of e.*
9. *e is in front of d unless it (i.e., e) is a large tetrahedron.*
10. *At least one of a, c, and e is a cube.*
11. *a is a cube only if b is in front of c.*
12. *b is larger than both a and e.*
13. *a and e are both larger than c, but neither is large.*
14. *d is the same size as b.*
15. *a is large if and only if it's a cube.*
16. *e is a tetrahedron unless c is a cube.*
17. *If e isn't a tetrahedron, either b or d is small.*
18. *b or d is a dodecahedron if either a or c is a tetrahedron.*
19. *d is a dodecahedron just in case b is a cube.*
20. *b is a dodecahedron just in case c is.*

Save your list of sentences as Sentences 4.4.

Problem$^{\star}_{\diamond}$ 5. (Building a world) Build a world in which all of the English sentences listed in Exercise 4 are true. Now make sure that all your translations are also true. If one of your translations is false, see whether the original English sentence is true. If it is, then there is something wrong with your translation. Play the game to try to figure out what the problem is. Save your world as World 4.5.

Exercise$_{\diamond}$ 6. (Checking your translations) Open Bolzano's World. Notice that all the English sentences from Exercise 4 are true in it. Thus, if your translations are accurate, they will also be true in this world. Check to see that they are. If you made any mistakes, go back and fix them.

Remember that even if one of your sentences comes out true in Bolzano's World, it does not mean that it is a proper translation of the corresponding English sentence. If the translation is correct, it will have the same truth value as the English sentence in *every* world. So let's check your translations in some other worlds.

Open Leibniz's World. The English sentences 3, 5, 8, 9, 11, 12, 13, and 15 are false here, the rest are true. Check to see that the same holds of your translations. If not, correct your translations.

Open Wittgenstein's World. Here we see that the English sentences 3, 5, 9, 12, 13, 14, 18, and 19 are false, while the rest are true. Check to see that the same holds of your translations.

Problem$^{\star}_{\diamond}$ 7. (Name that object) Open Sherlock's World and Sherlock's Sentences. You will notice that none of the objects in this world has a name. Your task is to assign the names a, b, and c in such a way that all the sentences in the list come out true. Save the resulting world in a file named World 4.7.

Problem$^{\star}_{\diamond}$ 8. (Figuring out sizes and shapes) Start a new sentence file and use it to translate the following English sentences.

1. *If a is a tetrahedron, then b is also a tetrahedron.*
2. *c is a tetrahedron if b is.*
3. *a and c are both tetrahedra only if at least one of them is large.*
4. *a is a tetrahedron but c isn't large.*
5. *If d is a dodecahedron, then it is neither large nor small, assuming that c is small.*
6. *c is medium only if none of d, e, and f are cubes.*
7. *d is a small dodecahedron unless a is small.*
8. *e is large just in case it is a fact that d is large if and only if f is.*
9. *d and e are the same size.*
10. *d and e are the same shape.*

11. *f is either a cube or a dodecahedron, if it is large.*
12. *c is larger than e only if b is larger than c.*

Save these sentences as Sentences 4.8. Assume that all of these sentences are true in some world. Figure out the sizes and shapes of a, b, c, d, e, and f. Add to the list of translations six additional sentences stating your conclusions about their sizes and shapes. (Don't forget to save the expanded sentence file.) Then build a world in which these six sentences are true. Verify that all of your translations are true as well. [Hint: If you have trouble, open Boole's World. All the English sentences are true in this world.]

Problem 9. Using the symbols introduced in Table 1, page 23, translate the following sentences into first-order logic.

1. *If Claire gave Folly to Max at 2:03 then it belonged to her at 2:00 and to him at 2:05.*
2. *Max erased Folly at 2:00 pm, but if he gave it to Claire then, it was not blank five minutes later.*
3. *If neither Max nor Claire erased Folly at 2:00, then it wasn't blank.*
4. *Max was angry at 2:05 only if Claire erased either Folly or Silly five minutes before.*
5. *Max is a student if and only if Claire is not.*

Problem 10. Using Table 1 on page 23, translate the following into colloquial English.

1. (Erased(Max, Folly, 2:00) ∨ Erased(Claire, Folly, 2:00)) → Disk(Folly)
2. Erased(Max, Folly, 2:30) ↔ Erased(Claire, Silly, 2:00)
3. ¬Blank(Folly, 2:00) → Blank(Silly, 2:00)
4. ¬(Blank(Folly, 2:00) → Blank(Silly, 2:00))

Problem 11. Translate the following into FOL as best you can. Explain any predicates and function symbols you use, and any shortcomings in your first-order translations.

1. *If Abe can fool Stephen, surely he can fool Ulysses.*
2. *If you scratch my back, I'll scratch yours.*
3. *France will sign the treaty only if Germany does.*
4. *If Tweedledee gets a party, so will Tweedledum, and vice versa.*
5. *If John and Mary went to the concert together, they must like each other.*

4.3 Conversational implicature

In translating from English to FOL, there are many problematic cases. For example, many students resist translating a sentence like *Max is at home unless Claire is at home* into:

$$\neg\mathsf{Home(Claire)} \rightarrow \mathsf{Home(Max)}$$

These students think that the meaning of this English sentence would be more accurately captured by the biconditional claim:

$$\neg\mathsf{Home(Claire)} \leftrightarrow \mathsf{Home(Max)}$$

The reason the latter seems natural is that when we assert the English sentence, there is some implication that if Claire *is* at home, then Max is *not*.

To resolve problematic cases like this, it is often useful to distinguish between the *truth conditions* of a sentence, and other things that in some sense follow from the assertion of the sentence. To take an obvious case, suppose someone asserts the sentence *It is a lovely day.* One thing you may conclude from this is that the speaker understands English. This is not part of what the speaker said, however, but part of what can be inferred from his saying it. The truth or falsity of the claim has nothing to do with his linguistic abilities.

The philosopher H. P. Grice developed a theory of what he called *conversational implicature* to help sort out the genuine truth conditions of a sentence from other conclusions we may draw from its assertion. These other conclusions are what Grice called "implicatures." We won't go into this theory in detail, but knowing a little bit about it can be a great aid in translation, so we present an introduction to Grice's theory.

Suppose we have an English sentence *S* that someone asserts, and we are trying to decide whether a particular conclusion we draw is part of the meaning of *S* or, instead, one of its implicatures. Grice pointed out that if the conclusion *is* part of the meaning, then it cannot be "cancelled" by some further elaboration by the speaker. Thus, for example, the conclusion that Max is home is part of the meaning of an assertion of *Max and Claire are home*, so we can't cancel this conclusion by saying *Max and Claire are home, but Max isn't home.* We would simply be contradicting ourselves.

Contrast this with the speaker who said *It is a lovely day.* Here, if he had gone on to say (perhaps reading from a phrase book) *Do you speak any French?* then the suggestion that the speaker understands English is effectively cancelled.

A more illuminating use of Grice's cancellability test concerns the expression *either... or.* Recall that we claimed that this should be translated into FOL as an inclusive disjunction, using $\vee$. We can now see that the suggestion that this phrase expresses exclusive disjunction is generally just a conversational implicature. For example, if the waiter says *You can have either soup or salad*, there is a strong implicature that you cannot have both. But it is clear that this is just an implicature, since the waiter could, without contradicting himself, go on to say *And you can have both, if you want.* Had the original *either... or* expressed the exclusive disjunction, this would be like saying *You can have soup or salad but not both, and you can have both, if you want.*

Recall our earlier example involving the sentence *Max is at home unless Claire is at home.* There we denied that the correct translation was:

$$\neg\mathsf{Home}(\mathsf{Claire}) \leftrightarrow \mathsf{Home}(\mathsf{Max})$$

which is equivalent to the conjunction of the correct translation:

$$\neg\mathsf{Home}(\mathsf{Claire}) \rightarrow \mathsf{Home}(\mathsf{Max})$$

with the additional claim:

$$\mathsf{Home}(\mathsf{Claire}) \rightarrow \neg\mathsf{Home}(\mathsf{Max})$$

Is this second claim part of the meaning of the English sentence, or is it simply a conversational implicature? Grice's cancellability test shows that it is just an implicature. After all, it makes perfectly good sense for the speaker to go on to say *On the other hand, if Claire is at home, I have no idea where Max is.* This elaboration takes away the suggestion that if Claire is at home then Max is not.

Exercises and Problems

Problem 12. Suppose Claire asserts the sentence *Max managed to get Carl home.* Does this logically imply, or merely implicate, that it was hard to get Carl home? Justify your answer.

Problem 13. Suppose Max asserts the sentence *We can walk to the movie or we can drive.* Does his assertion logically imply, or merely implicate, that we cannot both walk and drive? How does this differ from the soup or salad example?

4.4 Methods of proof involving $\rightarrow$ and $\leftrightarrow$

Since $\rightarrow$ and $\leftrightarrow$ can be defined in terms of the connectives discussed earlier, we could always eliminate them from proofs by means of these definitions. Thus, for example, if we wanted to prove $\mathsf{Q} \rightarrow \mathsf{R}$ we could just prove $\neg\mathsf{Q} \vee \mathsf{R}$, and then use the definition. In practice, however, this is not a good idea. It is more natural and intuitive to use rules that involve these symbols directly, especially with $\rightarrow$.

Proof steps

When discussing proof techniques in the previous chapter, we distinguished between simple proof steps, and more important methods of proof. The same distinction serves us here.

The most common proof step involving $\rightarrow$ goes by the Latin name *modus ponens*, or by the English "conditional elimination." The rule says that if you have established $Q \rightarrow R$ and you have established Q then you can infer R. This rule is obviously valid, since if $Q \rightarrow R$ and Q are both true, then so must be R.

There is a similar proof step for the biconditional, since the biconditional is logically equivalent to a conjunction of two conditionals. If you have established $Q \leftrightarrow R$ or $R \leftrightarrow Q$, then if you can establish Q, you can infer R. This is called "biconditional elimination."

Remember

Let P and Q be any sentences of FOL.

1. Modus ponens: From $P \rightarrow Q$ and P infer Q.
2. Biconditional elimination: From P and either $P \leftrightarrow Q$ or $Q \leftrightarrow P$, infer Q.

There are a number of useful logical equivalences involving our two new symbols. One of the most important is known as the Law of Contraposition. It states that $P \rightarrow Q$ is logically equivalent to $\neg Q \rightarrow \neg P$. This latter conditional is known as the contrapositive of the original conditional.

Here are some logical equivalences to bear in mind, beginning with contraposition:

$$\begin{aligned} P \rightarrow Q &\quad\Leftrightarrow\quad \neg Q \rightarrow \neg P \\ P \rightarrow Q &\quad\Leftrightarrow\quad \neg P \vee Q \\ \neg(P \rightarrow Q) &\quad\Leftrightarrow\quad P \wedge \neg Q \\ P \leftrightarrow Q &\quad\Leftrightarrow\quad (P \rightarrow Q) \wedge (Q \rightarrow P) \\ P \leftrightarrow Q &\quad\Leftrightarrow\quad (P \wedge Q) \vee (\neg P \wedge \neg Q) \end{aligned}$$

Method of conditional proof

One of the most important methods of proof is that of conditional proof, a method that allows you to prove a conditional statement. Suppose you want to prove $Q \rightarrow R$. The method is to assume, as an additional premise, the antecedent Q. If, under this additional assumption, you are able to prove R, then conditional proof allows you to infer that $Q \rightarrow R$ follows from the original premises.

Let's first give a schematic example that uses both modus ponens and conditional proof. We will show that $A \rightarrow C$ is a logical consequence of the two premises $A \rightarrow B$ and $B \rightarrow C$. This may seem obvious, but let's do it anyway, just to show how the method works in a simple case.

Proof: We are given, as premises, $A \rightarrow B$ and $B \rightarrow C$. We want to prove $A \rightarrow C$. With an eye toward using the rule of conditional proof, let us assume, in addition, that A is true. Then, by applying modus ponens using our first premise, we get B. Using modus ponens again, this time with our second premise, we get C. So we have established the consequent of our desired conditional on the basis of our assumption of A. Thus, by the rule of conditional proof, we know that A $\rightarrow$ C follows from the initial premises alone.

Let us now look at a somewhat more interesting example. This example will use both conditional proof and proof by contradiction. Let us prove:

$$\mathsf{Even}(\mathsf{n}^2) \rightarrow \mathsf{Even}(\mathsf{n})$$

Proof: The method of conditional proof tells us that we can proceed by assuming $\mathsf{Even}(\mathsf{n}^2)$ and proving $\mathsf{Even}(\mathsf{n})$. So assume that n^2 is even. To prove that n is even, we will use proof by contradiction. Thus assume that n is not even, that is, that it is odd. Then we can express n as $2m + 1$, for some m. But then we see that:

$$\begin{aligned} n^2 &= (2m+1)^2 \\ &= 4m^2 + 4m + 1 \\ &= 2(2m^2 + 2m) + 1 \end{aligned}$$

This shows that n^2 is odd, contradicting our first assumption. This contradiction shows that n is not odd, i.e., that it is even. Thus, by conditional proof, we have established $\mathsf{Even}(\mathsf{n}^2) \rightarrow \mathsf{Even}(\mathsf{n})$.

When proving conditional statements of the form P $\rightarrow$ Q, it is often much easier to prove the contrapositive $\neg$Q $\rightarrow$ $\neg$P, which is, as we have noted, logically equivalent. In such cases, we use conditional proof, but we assume $\neg$Q and try to prove $\neg$P. So, for example, if we want to prove

$$\mathsf{Irrational}(\mathsf{x}) \rightarrow \mathsf{Irrational}(\sqrt{\mathsf{x}})$$

we could assume that $\sqrt{x}$ is rational, and then show that x must be rational. This turns out to be easy. We ask you to do it in Problem 17.

Remember

The method of conditional proof: To prove P $\rightarrow$ Q, assume P and prove Q.

We can also use this method to prove biconditional statements. To prove Q $\leftrightarrow$ R by conditional proof, you need to do two things: assume Q and prove R; then assume R and prove Q.

There is one more form of proof involving $\leftrightarrow$ that is quite common in mathematics. Mathematicians are quite fond of finding results which show that a number of different conditions are equivalent. Thus you will find theorems stated in the following form. "The following conditions are all equivalent: Q_1, Q_2, Q_3."

In FOL, one would be forced to express this by means of three biconditionals.

$$Q_1 \leftrightarrow Q_2$$
$$Q_2 \leftrightarrow Q_3$$
$$Q_1 \leftrightarrow Q_3$$

This would involve six conditional proofs, two for each biconditional. However, we can cut our work in half by noting that it suffices to prove some cycle of results like the following:

$$Q_1 \rightarrow Q_2$$
$$Q_2 \rightarrow Q_3$$
$$Q_3 \rightarrow Q_1$$

These would be shown by three conditional proofs, rather than the six that would otherwise be required. When there are more than three statements to be proved the saving is even more dramatic.

Let's give a very simple example. We will prove that the following conditions on a natural number n are all equivalent:

1. n is even
2. n^2 is even
3. n^2 is divisible by 4.

Proof: Rather than prove all six biconditional, we prove that (3) → (2) → (1) → (3). Assume (3). Now clearly, if n^2 is divisible by 4, then it is divisible by 2, so we have (3) → (2). Next, we prove (2) → (1) by proving its contrapositive. Thus assume n is odd and prove n^2 is odd. Since n is odd, we can write it in the form $2k+1$. But then $n^2 = 4k^2 + 4k + 1$ which is also odd. Finally, let us prove (1) → (3). But this is obvious.

In general, when we apply this method, one looks for simple or obvious implications, like (1) → (3) above, and tries to build them into your cycle of conditionals.

Exercises and Problems

Problem$_\diamond$ 14. Open Conditional Sentences. Suppose that the sentences in this file are your premises. Now consider the five sentences listed below. Some of them are logical consequences of the sentences in Conditional Sentences, but some are not. For those that are consequences, give informal proofs. For those that are not consequences, build worlds where the premises are true but the conclusion false. Save the worlds as World 4.14.x, where x is the number of the sentence below.

1. Tet(e)
2. Tet(c) → Tet(e)
3. Tet(c) → Larger(f, e)
4. Tet(c) → LeftOf(c, f)
5. Dodec(e) → Smaller(e, f)

Problem 15. Prove: If $n + m$ is odd then $n \times m$ is even. (Compare with Problem 37 in Chapter 3.)

Problem 16. (Some standard valid steps in reasoning.) Give proofs of the following valid steps.

1. Modus Tollens: From A → B and ¬B infer ¬A.
2. Strengthening the Antecedent: From B → C infer (A ∧ B) → C.
3. Weakening the Consequent: From A → B infer A → (B ∨ C).
4. Constructive Dilemma: From A ∨ B, A → C, and B → D infer C ∨ D.
5. Transitivity of the biconditional: From A ↔ B and B ↔ C infer A ↔ C.

Problem 17. Prove Irrational(x) → Irrational($\sqrt{x}$).

Problem 18. Does it follow from the following premises that the unicorn is mythical? That it is magical? Justify your answer.

1. The unicorn, if it is not mythical, is a mammal, but if it is mythical, it is immortal.
2. If the unicorn is either immortal or a mammal, it is horned.
3. The unicorn, if horned, is magical.

Problem$_\diamond$ 19. Open Between Sentences. Determine whether this set of sentences is satisfiable or not. If it is, build a world where all the sentences are true, and save it as World 4.19. If not, give a proof that the sentences are contradictory. That is, assume all of them and derive a contradiction.

Problem 20. Give proofs of the following:

1. A → (B → A) is a logical truth.
2. (A → (B → C)) ↔ ((A ∧ B) → C) is a logical truth.
3. A ↔ ¬B is a logical consequence of A ∨ B ∨ C, B → (A → ¬C), and A ↔ C.
4. C ∧ D is a consequence of A ∨ (B ∧ C), ¬E, (A ∨ B) → (D ∨ E), and ¬A.
5. C → B is a consequence of ¬A → B, C → (D ∨ E), D → ¬C, and A → ¬E.

Problem 21. Prove that the following conditions on the natural number n are all equivalent. Use as few conditional proofs as possible.

1. n is divisible by 3
2. n^2 is divisible by 3
3. n^2 is divisible by 9
4. n^3 is divisible by 3
5. n^3 is divisible by 9
6. n^3 is divisible by 27

Problem 22. Consider the following sentences.

1. *If Max owned Folly at 2:00, then Claire owned Silly at the same time.*
2. *Claire owned either Silly or Folly at 2:00.*

Using the symbols introduced in Table 1, page 23, translate these sentences into FOL. Does either of them follow from the other? In both cases, either give an informal proof, or describe circumstances which make the premise true and the conclusion false. (You may assume that it is part of the meaning of "owned" that two people cannot own the same disk at the same time.)

4.5 Formal proofs

Let us now turn to the formal analogues of the methods of proof using the conditional and biconditional. Again, we have both introduction and elimination rules for each connective.

Rules for the conditional

The rule of *modus ponens* or conditional elimination is easily formalized. If you have proven both P → Q and P then you can assert Q, citing as justification these two earlier steps. It does not make any difference in which order the earlier steps appear. Schematically:

Conditional Elimination (→ Elim):

```
  | P → Q
  |   ⋮
  | P
  |   ⋮
▷ | Q
```

The corresponding introduction rule is the formal counterpart of the method of conditional proof. It involves the construction of a subproof. To prove a statement of the form P → Q we begin our subproof with the assumption of P and try to prove Q. If we succeed, then we are allowed to close off the subproof and conclude our desired conditional, citing the subproof as justification. Schematically:

Conditional Introduction ($\rightarrow$ Intro):

| P
| ⋮
| Q
▷ P $\rightarrow$ Q

Here is a simple example that involves both these rules, a formal proof of A $\rightarrow$ C from (A $\lor$ B) $\rightarrow$ C.

Step	Sentence	Justification
1.	(A $\lor$ B) $\rightarrow$ C	
2.	A	
3.	A $\lor$ B	$\lor$ Intro: 2
4.	C	$\rightarrow$ Elim: 1, 3
5.	A $\rightarrow$ C	$\rightarrow$ Intro: 2–4

Once we have conditional introduction at our disposal, we can convert any proof with premises into the proof, without premises, of a corresponding conditional. For example, we showed in the last chapter how to give a formal proof of $\neg\neg$A from premise A. Using this, we can give a proof of the logically true sentence A $\rightarrow$ $\neg\neg$A.

Step	Sentence	Justification
1.	A	
2.	$\neg$A	
3.	A $\land$ $\neg$A	$\land$ Intro: 1, 2
4.	$\neg\neg$A	$\neg$ Intro: 2–3
5.	A $\rightarrow$ $\neg\neg$A	$\rightarrow$ Intro: 1–4

Notice that the subproof here is identical to the original proof given on page 72. We simply took that proof and turned it into a proof, using conditional introduction, of A $\rightarrow$ $\neg\neg$A.

Rules for the biconditional

The rules for the biconditional are just what you would expect, given the rules for the conditional.

The elimination rule for the biconditional can be stated schematically as follows:

Biconditional Elimination ($\leftrightarrow$ Elim):

$P \leftrightarrow Q$ (or $Q \leftrightarrow P$)
⋮
P
⋮
▷ Q

This means that one can conclude Q if one can establish P and either of the biconditionals indicated.

The introduction rule for the biconditional $P \leftrightarrow Q$ requires that you give two subproofs, one showing that Q follows from P, and one showing that P follows from Q:

Biconditional Introduction ($\leftrightarrow$ Intro):

| P
| ⋮
| Q
| Q
| ⋮
| P
▷ $P \leftrightarrow Q$

Here is an example of a proof using biconditional introduction. It shows how to derive the first DeMorgan law within the system $\mathcal{F}$.

1. $\neg(P \wedge Q)$	
2. $\neg P \vee \neg Q$	Prev Thm (page 75): 1
3. $\neg P \vee \neg Q$	
4. $\neg(P \wedge Q)$	Prev Thm (page 78): 3
5. $\neg(P \wedge Q) \leftrightarrow (\neg P \vee \neg Q)$	$\leftrightarrow$ Intro: 1–2, 3–4

Substitution of equivalents

When discussing informal methods of proof, we noted that you could replace a sentence with any logically equivalent sentence, even when the sentence was a part of a larger whole. So for example the principle of double negation justifies not only the move from B to $\neg\neg B$, but also the move from, say, $A \rightarrow (B \vee D)$ to $A \rightarrow (\neg\neg B \vee D)$. The question arises as to whether this is, in general, a legitimate move within the system $\mathcal{F}$ as we have defined it.

It turns out that such substitutions can always be justified within $\mathcal{F}$. For instance, here is a proof that $A \rightarrow (\neg\neg B \lor D)$ follows from $A \rightarrow (B \lor D)$.

1. $A \rightarrow (B \lor D)$	
2. A	
3. $B \lor D$	$\rightarrow$ Elim: 1, 2
4. B	
5. $\neg\neg B$	Prev Thm: 4
6. $\neg\neg B \lor D$	$\lor$ Intro: 5
7. D	
8. $\neg\neg B \lor D$	$\lor$ Intro: 7
9. $\neg\neg B \lor D$	$\lor$ Elim: 3, 4–6, 7–8
10. $A \rightarrow (\neg\neg B \lor D)$	$\rightarrow$ Intro: 2–9

Although such proofs are always possible, they can be tedious. In the next section, we will introduce an extended Fitch-style system that incorporates an explicit rule allowing the substitution of logical equivalents wherever they occur.

Exercises and Problems

Problem 23. Give formal proofs corresponding to the valid steps discussed in Problem 16, page 103.

Problem 24. Give formal proofs of the following:

1. $A \rightarrow (B \rightarrow A)$ from no premises
2. $(A \rightarrow (B \rightarrow C)) \leftrightarrow ((A \land B) \rightarrow C)$ from no premises
3. $C \land D$ from premises $A \lor (B \land C)$, $\neg E$, $(A \lor B) \rightarrow (D \lor E)$, and $\neg A$

Problem 25. Give formal proofs of the following important equivalences. Once you have proven these, you will frequently find it useful to cite them.

1. $(P \rightarrow Q) \leftrightarrow (\neg P \lor Q)$.
2. $\neg(P \rightarrow Q) \leftrightarrow (P \land \neg Q)$.

Problem 26. Give a formal proof of Cube(a) $\leftrightarrow$ Small(a) from the following set of premises:

1. Cube(a) $\lor$ Dodec(a) $\lor$ Tet(a)
2. Small(a) $\lor$ Medium(a) $\lor$ Large(a)
3. Medium(a) $\leftrightarrow$ Dodec(a)
4. Tet(a) $\leftrightarrow$ Large(a)

You will need to use the more general construal of contradiction discussed on page 72.

4.6 $\mathcal{F}'$: Speeding up system $\mathcal{F}$

The Fitch system $\mathcal{F}$ as we have presented it corresponds to the standard rules for propositional logic. There is a sense (discussed in the last section of this book) in which this system is strong enough to prove all logical truths of propositional logic. However, as you have no doubt discovered, the system can be rather painful to use. Things which seem obvious can be extremely tedious to prove.

In this section we add some new rules to the system. We call the resulting system $\mathcal{F}'$, read "F prime." These new rules do not allow us to prove anything new, but they do allow us to prove things in more natural ways in $\mathcal{F}'$.

Generalized substitution

The first rule we add to $\mathcal{F}'$ is a generalization of biconditional elimination. It will allow us, among other things, to substitute logically equivalent sentences for one another, even when they appear within a larger context. To state the rule, we write S(P) for a sentence that contains the sentence P as a part, and S(Q) for the result of substituting Q for P wherever it occurs in that sentence. Schematically, the rule looks like this:

Generalized Substitution (Gen Sub):

| P ↔ Q (or Q ↔ P)
| ⋮
| S(P)
| ⋮
▷ | S(Q)

Notice that biconditional elimination is just a special case of this rule—the case where S(P) is just P itself. But here is a simple example in which generalized substitution lets us do something that biconditional elimination would not.

| 1. A ↔ B
| 2. ¬B ∨ D
|—
| 3. ¬A ∨ D   Gen Sub: 1, 2

When we say that biconditional elimination would not let us do this, we do not mean that we could not prove the validity of this argument in system $\mathcal{F}$. But it would be a considerably longer proof, and would require the use of several other rules besides biconditional elimination.

One of the more convenient uses of generalized substitution is in replacing parts of a sentence with logically equivalent parts. Let's look at one example of how this works.

1. $\neg(\neg P \lor \neg Q)$	
2. $\neg(P \land Q) \leftrightarrow (\neg P \lor \neg Q)$	Prev Thm (page 106)
3. $\neg\neg(P \land Q)$	Gen Sub: 2, 1
4. $P \land Q$	$\neg$ Elim: 3

Generalized negations

Suppose we have two sentences, say P and $\overline{P}$, which differ only in that the latter results from adding to or deleting from the front of P an odd number of negation symbols. Then it is clear by the laws of double negation that $\overline{P}$ is logically equivalent to $\neg P$. Our second streamlining of the system $\mathcal{F}$ is to treat any such $\overline{P}$ and $\neg P$ as interchangeable in the negation introduction and elimination rules.

Suppose, for example, that we have established a contradiction in a subproof that begins with P. Then system $\mathcal{F}$ allows us to use the rule of negation introduction to assert $\neg P$. But in our new system $\mathcal{F}'$, we are allowed to assert any sentence $\overline{P}$ of the above form, one that results by adding to or deleting from the front of P an odd number of negations. So, for example, if P is the sentence ¬Cube(a), then our new system will allow us to assert Cube(a) directly, without the detour through ¬¬Cube(a) demanded by system $\mathcal{F}$.

Our modified negation rules take the following form:

Negation Elimination (¬ Elim):

$\neg\overline{P}$
⋮
▷ P

Negation Introduction (¬ Intro):

| P
| ⋮
| $Q \land \overline{Q}$
▷ $\overline{P}$

We note that the terms "introduction" and "elimination" are no longer quite appropriate for these rules since, for example, ¬¬¬P can be inferred from $\neg$P by applying negation elimination. But we won't let that oddity bother us.

Contradictions

Our next move is to introduce a new symbol $\bot$ (usually called "bottom") in proofs in $\mathcal{F}'$. Informally, we think of $\bot$ as a stand-in for any contradiction. There are two rules of proof for it: if you prove a contradiction, then you can assert $\bot$, and if you have established $\bot$, then you can assert anything at all. Schematically:

Bottom Introduction ($\bot$ Intro):

```
  | P
  | ⋮
  | ¬P
  | ⋮
▷ | ⊥
```

Bottom Elimination ($\bot$ Elim):

```
  | ⊥
  | ⋮
▷ | P
```

These rules do not allow us to prove anything we could not prove before, but they make many proofs much more natural. Here, for example, is a more natural proof of $\neg$(P $\land$ Q) from $\neg$P $\lor$ $\neg$Q than we were able to give on page 78. Notice how closely it corresponds to the informal proof of this result given on page 76.

```
| 1. ¬P ∨ ¬Q
|-----
|  | 2. P ∧ Q
|  |-----
|  |  | 3. ¬P
|  |  |-----
|  |  | 4. P                  ∧ Elim: 2
|  |  | 5. ⊥                  ⊥ Intro: 4, 3
|  |
|  |  | 6. ¬Q
|  |  |-----
|  |  | 7. Q                  ∧ Elim: 2
|  |  | 8. ⊥                  ⊥ Intro: 7, 6
|  | 9. ⊥                     ∨ Elim: 1, 3–5, 6–8
| 10. ¬(P ∧ Q)                ¬ Intro: 2–9
```

Subproofs with multiple premises

One striking feature of the system $\mathcal{F}$ is that proofs can have multiple premises, but there is no way to use such proofs as subproofs of other proofs. Subproofs always have exactly one premise when they appear in the rules of $\mathcal{F}$.

There are four rules in $\mathcal{F}$ that use subproofs: disjunction elimination, negation introduction, conditional introduction, and biconditional introduction. Each of these rules can be generalized in ways that allow multiple premises. The most useful of these is the generalization of negation introduction, so we begin with it. We leave the other generalizations to the problems that follow.

General proof by contradiction

Suppose we have a proof with multiple premises, say $\mathsf{P}_1, \ldots, \mathsf{P}_n$, that results in a contradiction. Then we know that one of the premises must fail. So we should be allowed to assert the disjunction of their denials. More generally, using our convention about $\overline{\mathsf{P}}$, we can state the rule as follows:

Proof by Contradiction (Contra):

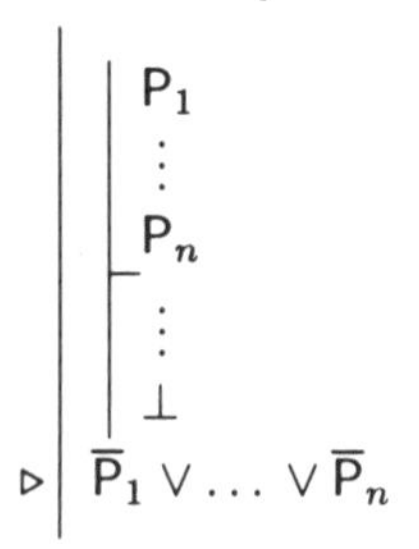

A striking example of the use of these three rules comes in the proof of $\mathsf{P} \vee \neg\mathsf{P}$. Recall the proof we gave of this on page 79, and compare it with the following:

1. $\neg\mathsf{P}$	
2. P	
3. $\perp$	$\perp$ Intro: 1, 2
4. $\mathsf{P} \vee \neg\mathsf{P}$	Contra: 1–3

Notice that in step 4, we have used our generalized negations to strip a negation off the first assumption of the subproof, but to add one to the second assumption.

For another example, here is a proof of $\neg\mathsf{P} \vee \neg\mathsf{R}$ from $\neg(\mathsf{P} \wedge \mathsf{R})$ in our enriched system $\mathcal{F}'$. This six line proof should be compared with the sixteen line proof on page 75.

1. $\neg(P \wedge R)$	
2. P	
3. R	
4. $P \wedge R$	$\wedge$ Intro: 2, 3
5. $\bot$	$\bot$ Intro: 4, 1
6. $\neg P \vee \neg R$	Contra: 2–5

Exercises and Problems

Problem 27. We can also modify the rule of conditional introduction to allow the subproof to have multiple assumptions. This modified rule makes it natural to prove a conditional whose antecedent is a conjunction. Give a statement of such a rule and use it to prove $(A \wedge B) \rightarrow C$ from $A \rightarrow C$.

Problem 28. We can modify the rule of disjunction elimination to allow us to have subproofs with multiple premises. Consider how you would like to proceed in cases where you have established a disjunction of conjunctions—a sentence in disjunctive normal form, for example. Give a statement of a valid rule that would be useful in such a case. Use it to give a short proof of B from $(A \wedge B) \vee (C \wedge B)$.

Problem 29. Give formal proofs of the following in system $\mathcal{F}'$.

1. $A \leftrightarrow \neg B$ from premises $A \vee B \vee C$, $B \rightarrow (A \rightarrow \neg C)$, and $A \leftrightarrow C$.
2. $C \rightarrow B$ from premises $\neg A \rightarrow B$, $C \rightarrow (D \vee E)$, $D \rightarrow \neg C$, and $A \rightarrow \neg E$.

4.7 Alternative notation

As with the other truth functional connectives, there are alternative notations for the material conditional and biconditional. The most common alternative to $P \rightarrow Q$ is $P \supset Q$. Polish notation for the conditional is Cpq. The most common alternative to $P \leftrightarrow Q$ is $P \equiv Q$. The Polish notation for the biconditional is Epq.

Remember

The following table summarizes the alternative notations discussed so far.

Our notation	Common equivalents
$\neg P$	$\sim P$, $\overline{P}$, $!P$, Np
$P \wedge Q$	$P\&Q$, $P\&\&Q$, $P \cdot Q$, PQ, Kpq
$P \vee Q$	$P \mid Q$, $P \parallel Q$, Apq
$P \rightarrow Q$	$P \supset Q$, Cpq
$P \leftrightarrow Q$	$P \equiv Q$, Epq

Part II

Quantifiers

5

Introduction to Quantification

Much of the expressive power of FOL stems not from the truth functional connectives, but from the so-called "quantifier" symbols, $\forall$ (read "for all") and $\exists$ ("there exists"). But before we can introduce these symbols, we need to introduce a new type of term, namely variables.

5.1 Variables

Variables are a kind of auxiliary symbol. In some ways they behave like individual constants, since they can appear in the list of arguments immediately following a predicate. But in other ways they are very different from individual constants. In particular, their semantic function is not to refer to objects. Rather, they are placeholders that indicate relationships between quantifiers and the argument positions of various predicates. This will become clearer with our discussion of quantifiers. First-order logic assumes an infinite list of variables so that we will never run out of them, no matter how complex a sentence may get. But Tarski's World uses only six variables, namely, u, v, w, x, y, and z. This imposes an expressive limitation on the language used in Tarski's World, but in practice one rarely has call for more than four or five variables.

5.2 Atomic wffs

Now that we have variables at our disposal, we can produce expressions that look like atomic sentences, except that there are variables in place of some individual constants. For example, Home(x) and Taller(John,y) are such expressions. We call them *atomic well-formed formulas*, or *atomic wffs*. They are not sentences, but will be used in conjunction with quantifier symbols to build sentences. The term "sentence" is reserved for well-formed formulas in which any variables that do occur are used together with quantifiers which "bind" them. We will give the definitions of sentence and bound variable in due course.

5.3 Quantifiers

Our language contains two quantifier symbols, ∀ and ∃. The reason these are called "quantifiers" is that they can be used to express certain rudimentary claims about the number (or quantity) of things that satisfy some condition. Specifically, they allow us to say that *all* objects satisfy some condition, or that *at least one* object satisfies some condition. When used in conjunction with identity (=) and the various connectives, they can also be used to express more complex numerical claims, for instance, that there are exactly two things that satisfy some condition.

Universal quantifier (∀)

This symbol is used to express universal claims, those we express in English using such terms as *everything, each thing, all things,* and *anything.* It is always used in connection with one of the variables u, v, w, x, ..., and so is said to be a variable binding operator. The combination ∀x is read "for every object x," or (somewhat misleadingly) "for all x."[1] If we wanted to translate the (rather unlikely) English sentence *Everything is at home* into first-order logic, we would use the expression

$$\forall x\ \mathsf{Home}(x)$$

This says that every object x meets the following condition: x is at home. Or, to put it more naturally, it says that everything whatsoever is at home.

Of course we rarely make such unconditional claims about absolutely everything. More common are restricted universal claims like *Every doctor is smart.* This sentence would be translated as:

$$\forall x\ (\mathsf{Doctor}(x) \rightarrow \mathsf{Smart}(x))$$

This sentence claims that given any object at all—call it x—if x is a doctor, then x is smart. To put it another way, the sentence says that if you pick anything at all, you'll find either that it is not a doctor or that it is smart (or perhaps both).

Existential quantifier (∃)

This symbol is used to express existential claims, those we express in English using such terms as *something, at least one thing, a,* and *an.* It too is always used in connection with one of the variables u, v, w, x, ..., and so is a variable binding operator. The combination ∃x is read "for some object x," or (somewhat misleadingly) "for some x." If we wanted

[1]We encourage students to use the first locution when reading formulas, at least for a few weeks, since we have seen many students who have misunderstood the basic function of variables as a result of reading them the second way.

to translate the English sentence *Something is at home* into first-order logic, we would use the expression

$$\exists x\ \mathsf{Home}(x)$$

This says that some object x meets the following condition: x is at home.

While it is possible to make such claims, it is more common to assert that something of a particular kind meets some condition, say, *Some doctor is smart.* This sentence would be translated as:

$$\exists x\ (\mathsf{Doctor}(x) \land \mathsf{Smart}(x))$$

This sentence claims that some object, call it x, meets the complex condition: x is both a doctor and smart. Or, more colloquially, it says that there is at least one smart doctor.

5.4 Wffs and sentences

Notice that in some of the above examples, we formed *sentences* out of complex expressions that were not themselves sentences, expressions like

$$\mathsf{Doctor}(x) \land \mathsf{Smart}(x)$$

that contain variables not bound by any quantifier. Thus, to systematically describe all the sentences of first-order logic, it is convenient to first describe a larger class, the so-called well-formed formulas, or *wffs.*

We have already explained what an atomic is: any n-ary predicate followed by n variables or individual constants. Using these as our atomic building blocks, we can construct more complicated wffs by repeatedly applying the following rules:

1. If P is a wff, so is $\neg \mathsf{P}$
2. If $\mathsf{P}_1, \ldots, \mathsf{P}_n$ are wffs, so is $(\mathsf{P}_1 \land \ldots \land \mathsf{P}_n)$
3. If $\mathsf{P}_1, \ldots, \mathsf{P}_n$ are wffs, so is $(\mathsf{P}_1 \lor \ldots \lor \mathsf{P}_n)$
4. If P and Q are wffs, so is $(\mathsf{P} \rightarrow \mathsf{Q})$
5. If P and Q are wffs, so is $(\mathsf{P} \leftrightarrow \mathsf{Q})$
6. If P is a wff and ν is a variable (i.e., one of u, v, w, x, ...), then $\forall \nu \mathsf{P}$ is a wff, and any occurrence of ν in P is said to be bound.
7. If P is a wff and ν is a variable, then $\exists \nu \mathsf{P}$ is a wff, and any occurrence of ν is P in said to be bound.

The way these rules work is pretty straightforward. For example, starting from the atomic wffs Cube(x) and Small(x) we can apply rule 2 to get the wff:

$$(\mathsf{Cube}(x) \land \mathsf{Small}(x))$$

Similarly, starting from the atomic wff LeftOf(x, y) we can apply rule 7 to get the wff:

$$\exists y\ \mathsf{LeftOf}(x, y)$$

In this formula the variable y has been bound by the quantifier ∃y. The variable x, on the other hand, has not been bound; it is still "free."

The rules can also be applied to complex wffs, so from the above two wffs and rule 4 we can generate the following wff:

$$((\mathsf{Cube}(x) \land \mathsf{Small}(x)) \rightarrow \exists y\ \mathsf{LeftOf}(x, y))$$

A *sentence* is a wff with no unbound (free) variables. None of the wffs displayed above are sentences, since they all contain unbound variables. To get a sentence from the last of these, we can simply apply rule 6 to produce:

$$\forall x\ ((\mathsf{Cube}(x) \land \mathsf{Small}(x)) \rightarrow \exists y\ \mathsf{LeftOf}(x, y))$$

Here all occurrences of the variable x have been bound by the quantifier ∀x. So this wff is a sentence since it has no free variables. It claims that for every object x, if x is both a cube and small, then there is an object y such that x is to the left of y. Or, to put it more naturally, every small cube is to the left of something.

These rules can be applied over and over again to form more and more complex wffs. So, for example, repeated application of the first rule to the wff Home(John) will give us all of the following wffs:

$$\neg\mathsf{Home}(\mathsf{John})$$
$$\neg\neg\mathsf{Home}(\mathsf{John})$$
$$\neg\neg\neg\mathsf{Home}(\mathsf{John})$$
$$\vdots$$

Since none of these contains any variables, and so no free variables, they are all sentences. They claim, as you know, that John is not home, that it is not the case that John is not home, that it is not the case that it is not the case that John is not home, and so forth.

We have said that a sentence is a wff with no free variables. However, it can sometimes be a bit tricky deciding whether a variable is free in a wff. For example, there are no free variables in the wff,

$$\exists x\ (\mathsf{Doctor}(x) \land \mathsf{Smart}(x))$$

However there *is* a free variable in the deceptively similar wff,

$$\exists x\ \mathsf{Doctor}(x) \land \mathsf{Smart}(x)$$

Here the last occurrence of the variable x is still free. We can see why this is the case by thinking about when the existential quantifier was applied in building up these two formulas. In the first one, the parentheses show that the quantifier was applied to the conjunction (Doctor(x) ∧ Smart(x)). As a consequence, all occurrences of x in the conjunction were bound by this quantifier. In contrast, the lack of parentheses show that in building up the second formula, the existential quantifier was applied to form ∃x Doctor(x), thus binding only the occurrence of x in Doctor(x). This formula was then conjoined with Smart(x), and so the latter's occurrence of x did not get bound.

Parentheses, as you can see from this example, make a big difference. They are the way you can tell what the "scope" of a quantifier is, that is, which variables fall under its influence and which don't.

Exercises and Problems

Problem◊ 1. (Fixing some expressions) Open the sentence file Bernstein's Sentences. The expressions in this list are not quite well-formed sentences of our language, but they can all be made sentences by slight modification. Turn them into sentences *without adding or deleting any quantifier symbols.* With some of them, there is more than one way to make them a sentence. Use **Verify** to make sure your results are well-formed sentences. Save your results as Sentences 5.1.

Problem◊ 2. (Fixing some more expressions) Open the sentence file Schönfinkel's Sentences. Again, the expressions in this list are not well-formed sentences. Turn them into sentences, but this time, do it *only* by adding or deleting quantifier symbols (and variables, if necessary). Do not add any parentheses. Use **Verify** to make sure your results are well-formed sentences. Save your results as Sentences 5.2.

Problem◊ 3. (Fixing ungrammatical expressions) Open Bozo's Sentences and Leibniz's World. Some of the expressions in this sentence file are not wffs, while others are wffs but not sentences. Read and assess each one. If it is not a wff, fix it. If it's not a sentence, adjust it so as to make it a true sentence with as little change as possible. If it's a false sentence, try to make it true, again with as little change as possible. See if you can capture the intent of the original expression. Save your sentences as Sentences 5.3.

5.5 Semantics for the quantifiers

When we described the meanings of our various connectives, we told you how the truth value of a complex sentence, say ¬P, depends on the truth values of its constituents, in this case P. But we have not yet given you similar rules for determining the truth value of quantified sentences. The reason is simple: the expression to which we apply the quantifier in order to build a sentence is usually not itself a sentence. We could hardly tell you how the truth value of ∃x Cube(x) depends on the truth value of Cube(x), since this latter expression is not a sentence at all: it contains a free variable. Because of this, it is *neither* true *nor* false.

To describe when quantified sentences are true, we need to introduce the auxiliary notion of *satisfaction.* The basic idea is simple, and can be illustrated with a few examples. We say that an object satisfies the atomic wff Cube(x) if and only if the object is a cube. Similarly, we say an object satisfies the complex wff Cube(x) ∧ Small(x) if and only if it is both a cube and small. As a final example, an object satisfies the wff Cube(x) ∨ ¬Large(x) if and only if it is either a cube or not large (or both).

Different logic books treat satisfaction in somewhat different ways. We will describe the one that is built into the way that Tarski's World checks the truth of quantified sentences. Suppose S(x) is a wff containing x as its only free variable, and suppose we wanted to know whether a given object satisfies S(x). If this object has a name, say b, then form a new *sentence* S(b) by replacing all free occurrences of x by the individual constant b. If the new sentence S(b) is true, then the object satisfies the formula S(x); if the sentence is not true, then the object does not satisfy the formula.

This works fine as long as the given object has a name. However, first-order logic does not require that every object have a name. How can we define satisfaction for objects that don't have names? It is for this reason that Tarski's World has, in addition to the individual constants a, b, c, d, e, and f, a further list n1, n2, n3, ... of individual constants. If we want to know whether a certain object without a name satisfies the formula S(x), we choose the first of these individual constants not in use, say n7, temporarily name the given object with this symbol, and then check to see whether the sentence S(n7) is true. Thus any small cube satisfies Cube(x) ∧ Small(x), because if we were to use n7 as a name of such a small cube, then Cube(n7) ∧ Small(n7) would be a true sentence.

Once we have the notion of satisfaction, we can easily describe when a sentence of the form ∃x S(x) is true. It will be true if and only if there is at least one object that satisfies the constituent wff S(x). So ∃x (Cube(x) ∧ Small(x)) is true if there is at least one object that satisfies Cube(x) ∧ Small(x), that is, if there is at least one small cube. Similarly, a sentence of the form ∀x S(x) is true if and only if every object satisfies the constituent wff S(x). Thus ∀x (Cube(x) → Small(x)) is true if every object satisfies Cube(x) → Small(x), that is, if every object either isn't a cube or is small.

This approach to satisfaction is conceptually simpler than some. A more common approach is to avoid the introduction of new names by defining satisfaction for wffs with an arbitrary number of free variables. We will not need this for specifying the meaning of quantifiers, but we will need it in some of the more advanced sections. For this reason, we postpone the general discussion until later.

In giving the semantics for the quantifiers, we have implicitly assumed that there is a relatively clear collection of objects that we are talking about. For example, if we encounter the sentence ∀x Cube(x) in Tarski's World, we interpret this to be a claim about the objects depicted in the world window. We do not judge it to be false just because the moon is not a cube. Similarly, if we encounter the sentence $\forall x[\mathsf{Even}(x^2) \rightarrow \mathsf{Even}(x)]$, we interpret this as a statement about numbers. It is true because every object in the domain we are talking about, natural numbers, satisfies the constituent wff.

In general, sentences containing quantifiers are only true or false relative to some domain of discourse. Sometimes the intended domain contains all objects there are. Usually, though, the intended domain is a much more restricted collection of things, say the people in the room, or some particular set of physical objects, or some collection of numbers. In this book, we will specify the domain explicitly unless it is clear from context what domain is intended.

Remember

- Quantified sentences make claims about some intended domain of discourse.
- A sentence of the form ∀xS(x) is true if and only if the wff S(x) is satisfied by every object in the domain of discourse.
- A sentence of the form ∃xS(x) is true if and only if the wff S(x) is satisfied by some object in the domain of discourse.

5.6 Game rules for the quantifiers

The game rules for the quantifiers are more interesting than those for the truth-functional connectives. With the connectives, moves in the game involved choosing sentences that are a part of the sentence to which you are committed. With the quantifier rules, however, moves consist in choosing objects, not sentences.

Suppose, for example, that you are committed to the truth of ∃xP(x). This means that you are committed to there being an object that satisfies P(x). Tarski's World will ask you to live up to this commitment by finding such an object. On the other hand, if you are committed to the falsity of ∃xP(x), then you are committed to there being no object that satisfies P(x). In which case, Tarski's World gets to choose: it tries to find an object that does satisfy P(x), thus contradicting your commitment.

The rules for ∀ are just the opposite. If you are committed to the truth of ∀xP(x), then you are committed to every object satisfying P(x). Tarski's World will try to find an object not satisfying P(x), thus contradicting your commitment. If, however, you are committed to the

TABLE 3
Summary of the game rules

FORM	YOUR COMMITMENT	PLAYER TO MOVE	GOAL
$P \vee Q$	TRUE FALSE	you Tarski's World	Choose one of P, Q that is true.
$P \wedge Q$	TRUE FALSE	Tarski's World you	Choose one of P, Q that is false.
$\exists x P(x)$	TRUE FALSE	you Tarski's World	Choose some ***b*** that satisfies the wff $P(x)$.
$\forall x P(x)$	TRUE FALSE	Tarski's World you	Choose some ***b*** that does not satisfy $P(x)$.
$\neg P$	either	—	Replace $\neg P$ by P and switch commitment.
$P \rightarrow Q$	either	—	Replace $P \rightarrow Q$ by $\neg P \vee Q$ and keep commitment.
$P \leftrightarrow Q$	either	—	Replace $P \leftrightarrow Q$ by $(P \rightarrow Q) \wedge (Q \rightarrow P)$ and keep commitment.

falsity of $\forall x P(x)$, then you are committed to there being some object that does not satisfy $P(x)$. Tarski's World will ask you to live up to your commitment by finding such an object.

We have now seen all the game rules. So let's stop and survey them. We summarize them in Table 3.

Exercises and Problems

Exercise$_\diamond$ 4. (Evaluating sentences in a world) Open Peirce's World and Peirce's Sentences. There are 30 sentences in this file. Work through them, assessing their truth and playing the game when necessary. You may need to switch to the 2-D view for some of the sentences.

Exercise$_\diamond$ 5. (Evaluating sentences in a world) Open Leibniz's World and Zorn's Sentences. The sentences in this file contain both quantifiers and the identity symbol. Work through them, assessing their truth and playing the game when necessary.

5.7 The four Aristotelian forms

Long before FOL was codified, Aristotle studied the kinds of reasoning associated with noun phrases like *Every man*, *No man*, and *Some man*, expressions we would translate using our quantifier symbols. The four main sentence forms treated in Aristotle's logic were the following.

All P's are Q's
Some P's are Q's
No P's are Q's
Some P's are not Q's

We will begin by studying the first two of these forms, which we have already discussed to a certain extent. These forms are translated as follows. The form *All P's are Q's* is translated as:

$$\forall x\,(P(x) \rightarrow Q(x))$$

whereas the form *Some P's are Q's* is translated as:

$$\exists x\,(P(x) \wedge Q(x))$$

Beginning students are often tempted to translate the latter more like the former, say as:

$$\exists x\,(P(x) \rightarrow Q(x))$$

This is in fact an extremely unnatural sentence of first-order logic. It is meaningful, but it doesn't mean what you might think. It is true just in case there is an object which is either not a P or else is a Q, which is something quite different than saying that some P's are Q's. This point is driven home in Exercise 7.

The other two Aristotelian forms are translated similarly, but using a negation. In particular *No P's are Q's* is translated

$$\forall x(P(x) \rightarrow \neg Q(x))$$

Many students, and one of the authors, finds it more natural to use the following, logically equivalent sentence:

$$\neg\exists x(P(x) \wedge Q(x))$$

Both of these assert that nothing that is a P is also a Q.

The final of the four forms, *Some P's are not Q's*, is translated by

$$\exists x(P(x) \wedge \neg Q(x))$$

which says that there is something which is a P but not a Q.

There are a number of topics that traditionally go with the study of the logic associated with these forms. We will not discuss these explicitly here. Rather, our use for these forms is simply as examples of the very simplest sorts of sentences built using quantifiers.

Exercises and Problems

Problem$_\diamond$ 6. (Building a world) Open Aristotle's Sentences. Each of these sentences is of one of the four Aristotelian forms. Build a single world where all the sentences in the file are true. As you work through the sentences, you will find yourself successively modifying the world. Whenever you make a change in the world, you had better go back and check that you haven't made any of the earlier sentences false. Then, when you are finished, verify that all the sentences are really true. Save your world as World 5.6.

Exercise$_\diamond$ 7. (A common translation mistake) Open Edgar's Sentences and evaluate them in Edgar's World. Make sure you understand why each of them has the truth value it does. Which of them would be a good translation of *There is a tetrahedron that is large?* (Clearly this English sentence is false in Edgar's World, since there are no tetrahedra at all.) Which would be a good translation of *There is a cube between a and b?* Can you express in English the claim made by sentence 5? How about sentence 6?

Problem$_\diamond$ 8. (Name that object) Open Maigret's World and Maigret's Sentences. The object is to try to figure out which objects have names, and what they are. You should be able to figure this out from the sentences, all of which are true. Once you have come to your conclusion, add names to the objects in the world and see if all the sentences do indeed evaluate as true. Save your modified world as World 5.8.

Problem$_\diamond$ 9. (Describing a world) Open Reichenbach's World 1. Start a new sentence file where you will describe some features of this world using sentences of the simple Aristotelian forms. Check each of your sentences to see that it is indeed a sentence and that it is true in this world.

1. Use your first sentence to describe the size of all the tetrahedra.
2. Use your second sentence to describe the size of all the cubes.
3. Use your third sentence to express the truism that every dodecahedron is either small, medium, or large.
4. Notice that some dodecahedron is large. Express this fact.
5. Observe that some dodecahedron is not large. Express this.
6. Notice that some dodecahedron is small. Express this fact.
7. Observe that some dodecahedron is not small. Express this.
8. Notice that some dodecahedron is neither large nor small. Express this.

9. Express the observation that no tetrahedron is large.
10. Express the fact that no cube is large.

Now change the sizes of the objects in the following way: make one of the cubes large, one of the tetrahedra medium, and all the dodecahedra small. With these changes, the following should come out false: 1, 2, 4, 7, 8, and 10. If not, then you have made an error in describing the original world. Can you figure out what it is? Try making other changes and see if your sentences have the expected truth values. Save your sentences as Sentences 5.9.

Problem 10. Assume we are working in an extension of the first-order language of arithmetic with the additional predicates Even(x) and Prime(x), meaning, respectively, "x is an even number" and "x is a prime number." Express the following in this language:

1. No even number is prime.
2. Every prime is either odd or equal to 2.
3. Some prime is even.
4. Some prime is not even.

Which of these sentences are true?

5.8 Translating complex noun phrases

The first thing you have to learn in order to translate quantified English expressions is how to treat complex noun phrases, expressions like "a boy living in Omaha" or "every girl living in Duluth." In this section we will learn how to do this. We concentrate first on the former sort of noun phrase, whose most natural translation involves an existential quantifier. Typically, these will be noun phrases starting with one of the determiners "some," "a," and "an," including noun phrases like "something." These are called existential noun phrases, since they assert the existence of something or other. Of course two of our four Aristotelian forms involve existential noun phrases, so we know the general pattern: existential noun phrases are usually translated using $\exists$, frequently together with $\wedge$.

Let's look at a simple example. Suppose we wanted to translate the sentence *A small, happy dog is at home.* This sentence claims that there is an object which is simultaneously a small, happy dog, and at home. We would translate it as:

$$\exists x[(\mathsf{Small}(x) \wedge \mathsf{Happy}(x) \wedge \mathsf{Dog}(x)) \wedge \mathsf{Home}(x)]$$

We have put parentheses around the first three predicates to indicate that they were all part of the translation of the subject noun phrase. But this is not really necessary.

Universal noun phrases are those that begin with determiners like "every," "each," and "all." These are usually translated with the universal quantifier. Sometimes noun phrases beginning with "no" and with "any" are also translated with the universal quantifier. Two of our four Aristotelian forms involve universal noun phrases, so we also know the general pattern here: universal noun phrases are usually translated using $\forall$, frequently together with $\rightarrow$.

Let's consider the sentence *Every small dog that is at home is happy.* This claims that everything with a complex property, that of being a small dog at home, has another property, that of being happy. This suggests that the overall sentence has the form *All A's are B's.* But in this case, to express the complex property that fills the "*A*" position, we will use a conjunction. Thus it would be translated as:

$$\forall x[(\mathsf{Small}(x) \land \mathsf{Dog}(x) \land \mathsf{Home}(x)) \rightarrow \mathsf{Happy}(x)]$$

In this case, the parentheses are not optional. Without them the expression would not be well formed.

You will be given lots of practice translating complex noun phrases in the problems that follow. First, however, we discuss some troublesome cases.

Conversational implicature and quantification

You will find that translating quantified phrases is not difficult, as long as quantifiers are not "nested" inside one another. There are, however, a couple of points that sometimes present stumbling blocks.

One thing that often puzzles students has to do with the truth value of sentences of the form

$$\forall x(\mathsf{P}(x) \rightarrow \mathsf{Q}(x))$$

in worlds where there are no objects satisfying $\mathsf{P}(x)$. If you think about it, or work Exercise 13, you will see that in such a world the sentence is true simply because there are no objects that satisfy the antecedent. This is called a *vacuously* true generalization.

Consider, for example, the sentence

$$\forall y(\mathsf{Tet}(y) \rightarrow \mathsf{Small}(y))$$

which asserts that every tetrahedron is small. But imagine that it has been asserted about a world in which there are no tetrahedra. In such a world the sentence is true simply because there are no tetrahedra at all, small, medium, or large. Consequently, it is impossible to find a counterexample, a tetrahedron which is not small.

What strikes students as especially odd are examples like

$$\forall y(\mathsf{Tet}(y) \rightarrow \mathsf{Cube}(y))$$

On the face of it, such a sentence looks contradictory. But we see that if it is asserted about a world in which there are no tetrahedra, then it is in fact *true*. But that is the only way it can be true: if there are no tetrahedra. In other words, the only way this sentence can be true is if it is vacuously true. Let's call generalizations with this property "inherently vacuous." Thus a sentence of the form $\forall x\,(P(x) \rightarrow Q(x))$ is inherently vacuous if any world in which it is true is also a world in which $\forall x\, \neg P(x)$ is true.

In everyday conversation, it is rare to encounter a vacuously true generalization. When we do, we feel that the speaker has misled us. For example, suppose a professor claims "Every freshman who took the class got an A," when in fact no freshman took her class. Here we wouldn't say that she lied, but we would certainly say that she misled us. Her statement typically carries the conversational implicature that there were freshmen in the class. If there were no freshmen, then that's what she would have said if she were being forthright. This is why *inherently* vacuous claims strike us as counterintuitive: we see that they cannot be true without being misleading.

Another source of puzzlement concerns the relationship between the following two Aristotelian sentences:

Some P's are Q's
All P's are Q's

Students often have the intuition that the first should contradict the second. After all, why would you say that *some* student got an A if *every* student got an A? If this intuition were right, then the correct translation of *Some P's are Q's* would not be what we have suggested above, but rather

$$\exists x(P(x) \land Q(x)) \land \neg\forall x(P(x) \rightarrow Q(x))$$

It is easy to see, however, that the second conjunct of this sentence does not represent part of the meaning of the sentence. It is, rather, another example of a conversational implicature. It makes perfectly good sense to say "Some student got an A on the exam. In fact, every student did." If the proposed conjunction were the right form of translation, this amplification would be contradictory.

Exercises and Problems

Problem$_\diamond$ 11. (Translating existential noun phrases)

- Start a new sentence file named Sentences 5.11 and enter translations of the following English sentences. Each will use the symbol $\exists$ exactly once. None will use the symbol $\forall$. As you go, check that your entries are well-formed sentences. By the way, you will find that many of these English sentences are translated using the same first-order sentence.

1. *Something is large.*
2. *Something is a cube.*
3. *Something is a large cube.*
4. *Some cube is large.*
5. *Some large cube is to the left of b.*
6. *A large cube is to the left of b.*
7. *b has a large cube to its left.*
8. *b is to the right of a large cube.* [Hint: This translation should be almost the same as the last, but it should contain the predicate symbol RightOf.]
9. *Something to the left of b is in back of c.*
10. *A large cube to the left of b is in back of c.*
11. *Some large cube is to the left of b and in back of c.*
12. *Some dodecahedron is not large.*
13. *Something is not a large dodecahedron.*
14. *It is not the case that something is a large dodecahedron.*
15. *b is not to the left of a cube.* [Warning: This sentence is ambiguous. Can you think of two importantly different translations? One starts with $\exists$, the other starts with $\neg$. Use the second of these for your translation, since this is the most natural reading of the English sentence.]

- Open Montague's World. Notice that all the English sentences above are true in this world. Check that all your translations are also true. If not, you have made a mistake. Can you figure out what is wrong with your translation?
- Move the large cube to the back right corner of the grid. Observe that English sentences 5, 6, 7, 8, 10, 11 and 15 are now false, while the rest are still true. Check that the same holds of your translations. If not, you have made a mistake. Can you figure out what is wrong with your translation?
- Now make the large cube small. The English sentences 1, 3, 4, 5, 6, 7, 8, 10, 11, and 15 are false in the modified world, the rest are true. Again, check that your translations have the same truth values. If not, figure out what is wrong.
- Finally, move c straight back to the back row, and make b large. All the English sentences other than 1, 2, and 13 are false. Check that the same holds for your translations. If not, figure out where you have gone wrong.

Problem$_\diamond$ 12. (Common mistakes, part 2) In this exercise we return to the point made in Exercise 7, page 124. Glance back at that exercise to recall the basic point. Now open and display all of Allan's Sentences.

1. Which of the sentences is a correct translation of *Some dodecahedron is large?* Which is a correct translation of *All tetrahedra are small?* If you are not sure, try constructing worlds of various sorts and testing the sentences to see when they come out true.
2. Construct a world in which sentences 2 and 4 are true, but sentences 1 and 3 are false. Save it as World 5.12.2.
3. Can you construct a world in which sentence 3 is true and sentence 4 is false? If so, do so and save it as World 5.12.3. If not, why not?
4. Can you construct a world in which sentence 1 is true and sentence 2 is false? If so, do so and save it as World 5.12.4. If not, why not?

Exercise$_\diamond$ 13. (Vacuously true generalizations) Open Dodgson's Sentences. Note that the first sentence says that every tetrahedron is large.

1. Open Peano's World. Sentence 1 is clearly false in this world, since the small tetrahedron is a "counterexample" to the universal claim. What this means is that if you play the game committed to the falsity of this claim, then when Tarski's World asks you to pick an object you will be able to pick the small tetrahedron and win the game. Try this.
2. Delete this counterexample and verify that sentence 1 is now true.
3. Now open Peirce's World. Verify that sentence 1 is again false, this time because there are three counterexamples. (Now if you play the game committed to the falsity of the sentence, you will have three different winning moves when asked to pick an object: you can pick any of the small tetrahedra and win.)
4. Delete all three counterexamples, and evaluate the claim. Is the result what you expected? The generalization is true, because there are no counterexamples to it. It is what we called a vacuously true generalization, since there are no objects that satisfy the antecedent. That is, there are no tetrahedra at all, small, medium, or large.
5. Confirm that all of sentences 1–3 are vacuously true in the current world.
6. Two more vacuously true sentences are given in sentences 4 and 5. However, these sentences are different in another respect. Each of the first three sentences could have been non-vacuously true in a world, but these latter two can only be true in worlds containing no tetrahedra. That is, they are inherently vacuous.

Problem$_\diamond$ 14. (Translating universal noun phrases)

- Start a new sentence file, and enter translations of the following sentences. This time each translation will contain exactly one $\forall$ and no $\exists$.

1. *All cubes are small.*
2. *Each small cube is to the right of a.*
3. *All dodecahedra are large.*
4. *a is to the left of every dodecahedron.*
5. *Every medium tetrahedron is in front of b.*
6. *Each cube is either in front of b or in back of a.*
7. *Every cube is to the right of a and to the left of b.*
8. *Everything between a and b is a cube.*
9. *Everything smaller than a is a cube.*
10. *All dodecahedra are not small.* [Note: Most people find this sentence ambiguous. Can you find both readings? One starts with ∀, the other with ¬. Use the former, the one that means, in effect, that all the dodecahedra are either medium or large.]
11. *No dodecahedron is small.*
12. *a is not to the right of everything.* [Note: This sentence is ambiguous. We want you to interpret it as a denial of the claim that a is to the right of everything.]
13. *a is not to the right of anything.* [Note: These last two sentences mean different things, though they can both be translated using ∀, ¬, and RightOf.]
14. *a is not to the right of any cube.*
15. (⋆) *If something is a cube, then it is right of a and left of b.* [Warning: While this sentence contains the noun phrase "something," it is actually making a universal claim, and so should be translated with ∀. You might first try to paraphrase it using the English phrase "every cube."]
16. (⋆) *Something is a cube if and only if it is right of a and left of b.*

Save your sentences as Sentences 5.14.

- Open Claire's World. Check to see that all the English sentences are true in this world. Now check to see that all your translations are true. Again, if you have made any mistakes, fix them.
- Adjust Claire's World by moving a to the front, right corner of the grid. With this change, the English sentences 2, 4, 7, 13, 14, 15, and 16 are false, while the rest are true. Make sure that the same holds of your translations. If not, try to figure out what is wrong and fix it.
- Next, open Wittgenstein's World. Observe that the English sentences 2, 4, 8, 9, 12, and 14 are true, but the rest are false. Check that the same is the case with your translations. If not, try to fix them.

- Finally, open Leibniz's World. The English sentences 3, 4, 8, 9, 10, 11, 12, 13, and 14 are true, but the rest are false. Check your translations.

Problem$_\diamond$ 15. (Translation) Open Leibniz's World. This time, we will translate some sentences while looking at the world they are meant to describe.

- Start a new sentence file, and enter translations of the following sentences. Each of the English sentences is true in this world. As you go, check to make sure that your translation is indeed a true sentence.

  1. *b is a tetrahedron and is smaller than e.*
  2. *There are no medium sized cubes.*
  3. *Nothing is in front of b.*
  4. *Every cube is either in front of or in back of e.*
  5. *No cube is between a and c.*

  Save your sentences as Sentences 5.15.

- Now let's change the world so that none of the English sentences is true. We can do this as follows. First change b into a medium cube. Next, delete the leftmost tetrahedron and move b to exactly the position just vacated by the late tetrahedron. Finally, add a small cube to the world, locating it exactly where b used to sit. If your answers to 1–5 are correct, all of the translations should now be false. Verify that they are.
- Make various changes to the world, so that some of the English sentences come out true and some come out false. Then check to see that the truth values of your translations track the truth values of the English sentences.

Problem 16. Using the symbols introduced in Table 1, page 23, translate the following into FOL.

1. *People are not disks.*
2. *Disks are not people.*
3. *Silly was not erased at either 2:00 or 2:05.*
4. *Claire erased Folly sometime between 2:00 and 3:00.*
5. *Claire gave a disk to each and every student at 2:00.*
6. *Claire had only blank disks at 2:00.*
7. *Of all the students, only Claire was angry at 3:00.*
8. *No one erased Folly at 2:00.*
9. *If anyone erased Silly at 2:00, then they were angry.*
10. *Whoever owned Silly at 2:00 was angry five minutes later.*

Problem 17. Using Table 1, page 23, translate the following into colloquial English.

1. $\forall$y(Person(y) $\rightarrow$ $\neg$Owned(y, Folly, 2:00))
2. $\neg\exists$x(Angry(x, 2:00) $\wedge$ Student(x) $\wedge$ Erased(x, Folly, 2:00))
3. $\forall$x((Person(x) $\wedge$ Gave(Max, Folly, x, 2:00)) $\rightarrow$ Angry(x, 2:05))
4. $\forall$t$\neg$Gave(Claire, Folly, Max, t)

Problem* 18. Translate the following into FOL, introducing names, predicates, and function symbols as needed. As usual, explain your predicates and function symbols, and any shortcomings in your translations.

1. *Only the brave know how to forgive.*
2. *No man is an island.*
3. *I care for nobody, not I,*
   If no one cares for me.
4. *Every nation has the government it deserves.*
5. *There are no certainties, save logic.*
6. *Misery (that is, a miserable person) loves company.*
7. *All that glitters is not gold.*
8. *There was a jolly miller once*
   Lived on the River Dee.
9. *If you praise everybody, you praise nobody.*
10. *Something is rotten in the state of Denmark.*

5.9 Logical equivalences involving negation and quantifiers

In Chapter 3 we learned the DeMorgan rules relating $\neg$ and the symbols $\wedge$ and $\vee$. These were useful in transforming complicated looking sentences into simple sentences. In particular, they were crucial in the study of the various normal forms that we looked at later in the chapter.

There are similar important rules relating $\neg$ and the quantifier symbols $\forall$ and $\exists$. This should not be surprising since one way of expressing $\forall$xP(x) would be to use a conjunction of the form

$$P(n_1) \wedge P(n_2) \wedge \ldots$$

where one simply names every object in the domain of discourse and asserts P of it. Clearly the one is true if and only if the other is. Similarly, one could treat $\exists$xP(x) as a long disjunction

$$P(n_1) \vee P(n_2) \vee \ldots$$

The trouble, of course, is that this may be impractical (say you are talking about grains of sand on the beach) or impossible (say the domain of discourse includes all real numbers). Still, intuitively, these are striking similarities. Thus, it is not surprising that we have logical equivalences which are entirely similar to the DeMorgan laws for the connectives.

Stated in English, the DeMorgan laws for the quantifiers allow you to "push a negation past a quantifier" by switching the quantifier. So, for example, if we know that not everything has some property ($\neg\forall x P(x)$), then we know that something does not have the property ($\exists x \neg P(x)$), and vice versa. Similarly, if we know that it is not the case that something has some property ($\neg\exists x P(x)$), then we know that everything must fail to have it ($\forall x \neg P(x)$), and vice versa.

Remember

(DeMorgan laws for quantifiers) For any wff $P(x)$:

1. $\neg\forall x P(x) \Leftrightarrow \exists x \neg P(x)$.
2. $\neg\exists x P(x) \Leftrightarrow \forall x \neg P(x)$.

Using these, we can see that there is a close relationship between certain pairs of Aristotelian sentences. In particular, the negation of *All P's are Q's* is logically equivalent to *Some P's are not Q's.* To see how this follows from DeMorgan's laws, we note the following chain of logical equivalences. The first is the translation of *It is not true that all P's are Q's* while the last is the translation of *Some P's are not Q's.*

$$\begin{aligned} \neg\forall x[P(x) \rightarrow Q(x)] &\Leftrightarrow \neg\forall x[\neg P(x) \lor Q(x)] \\ &\Leftrightarrow \exists x \neg[\neg P(x) \lor Q(x)] \\ &\Leftrightarrow \exists x[\neg\neg P(x) \land \neg Q(x)] \\ &\Leftrightarrow \exists x[P(x) \land \neg Q(x)] \end{aligned}$$

The first step uses the equivalence of $P(x) \rightarrow Q(x)$ and $\neg P(x) \lor Q(x)$. The second and third steps use DeMorgan's Laws, first one of the quantifier versions, and then one of the connective versions. The last step uses the double negation law applied to $\neg\neg P(x)$.

While we are giving principles governing the quantifiers, we should mention one that is so basic it can be easily overlooked. You will have noticed that when you are translating from English to FOL and find a quantified noun phrase, you must pick some variable to use in your translation, and that we have given you no guidelines as to which one is "right." This is because it does not matter which you use, as long as you pick a variable that is not already in use. We codify this by means of the following:

Remember

(Principle of Replacing Bound Variables) For any wff $P(x)$ and variable y that does not occur in $P(x)$:

1. $\forall x P(x) \Leftrightarrow \forall y P(y)$
2. $\exists x P(x) \Leftrightarrow \exists y P(y)$

Since logically equivalent sentences can be substituted for one another, according to our principle, the sentence

$$\mathsf{Cube(b)} \land \exists x(\mathsf{Dodec(x)} \land \mathsf{Larger(x, b)}) \land \forall x(\mathsf{Tet(x)} \rightarrow \mathsf{LeftOf(x, b)})$$

is logically equivalent to

$$\mathsf{Cube(b)} \land \exists y(\mathsf{Dodec(y)} \land \mathsf{Larger(y, b)}) \land \forall x(\mathsf{Tet(x)} \rightarrow \mathsf{LeftOf(x, b)})$$

Exercises and Problems

Problem 19. Use DeMorgan's Law to show that the negation of *No P's are Q's* is logically equivalent to *Some P's are Q's.*

Problem$_\diamond$ 20. (Using DeMorgan's Laws) Open DeMorgan's Sentences 2. Build any world at all and evaluate the first three sentences in that world. You should be able to predict the truth value of all the remaining sentences from these three. Use DeMorgan's Laws, the principle of replacing bound variables, and principles we have learned governing the connectives, to show that each of the remaining sentences is logically equivalent to one of the first three. Write out your proofs on paper.

Problem$_\diamond$ 21. ($\forall$ versus $\land$) We pointed out the similarity between $\forall$ and $\land$, as well as that between $\exists$ and $\lor$. We were careful not to claim that the universally quantified sentence was logically equivalent to the analogous conjunction. This problem will show you why we did not make this claim.

- Open Church's Sentences and Ramsey's World. Evaluate the sentences in this world. You will notice that the first two sentences have the same truth value, as do the second two.
- Modify Ramsey's World in any way you like, but do not add or delete objects, and do not change the names used. Verify that the first two sentences always have the same truth values, as do the last two.
- Now add one object to the world. Adjust the objects so that the first sentence is false, the second and third true, and the last false. Save your world as World 5.21. This shows that the first two sentences are not logically equivalent. Neither are the last two.

5.10 Methods of proof involving ∀ and ∃

In the preceding chapters we discussed valid patterns of reasoning that arise from the various connectives we have studied. It is now time to consider the valid patterns of reasoning associated with the quantifiers.

As with the connectives, there are both simple proof steps, and more substantive methods of proof. We will begin by discussing the simple proof steps that are most often used with ∀ and ∃.

Valid quantifier steps

Suppose that we are given as a premise, or have otherwise established, that everything in the domain of discourse is either a cube or a tetrahedron. And suppose we also know that c is in the domain of discourse. It follows, of course, that c is either a cube or a tetrahedron, since everything is.

More generally, if we have established ∀xS(x), and we know that c names an object in the domain of discourse, then we may legitimately infer S(c). After all, there is no way the universal claim could be true without the specific claim also being true. This inference step is called "universal instantiation" or "universal elimination." Notice that it allows you to move from a known result that begins with a quantifier ∀x(...x...) to one (...c...) where the quantifier has been eliminated.

There is a similar simple step for ∃, but it allows you to introduce the quantifier. Suppose, for example, we have established that c is a small tetrahedron. It follows, of course, that there is a small tetrahedron. There is no way for the specific claim about c to be true without the existential claim also being true. More generally, if we have established S(c) then we may infer ∃xS(x). This step is called "existential generalization" or "existential introduction."

In mathematical proofs, the preferred way to demonstrate the truth of an existential claim is to find (or construct) a specific instance that satisfies the requirement, and then apply existential generalization. For example, if we wanted to prove that there are natural numbers x, y, and z for which $x^2 + y^2 = z^2$, we could simply note that $3^2 + 4^2 = 5^2$ and apply existential generalization (thrice over).

The validity of these inference steps depends on a special convention of FOL that we have already stressed. The convention is that any name used denotes some object in the domain of discourse. English is a bit more subtle here. Consider, for example, the name *Santa*. The sentence

Claire believes Santa is coming

might be true in circumstances where one would be reluctant to conclude

There is someone who Claire believes is coming.

The trouble, of course, is that the name *Santa* does not denote anything. So one must be careful applying this rule in ordinary arguments if there are names in use that do not denote.

Remember

1. Universal instantiation: From $\forall x S(x)$, infer $S(c)$, so long as c denotes an object in the domain of discourse.
2. Existential generalization: From $S(c)$, infer $\exists x S(x)$, so long as c denotes an object in the domain of discourse.

We should say a bit about the restriction "so long as c denotes an object in the domain of discourse." In first-order logic, it is customary to assume that the domain of discourse includes any objects denoted by the names in use. So, for example, if both the names Claire and 2 are used in a proof, we must assume that the quantifiers range over both people and numbers. If we did not assume this, then extra caution in applying these rules would be required. The restriction is included in our statement of the rules simply to remind us of this.

Let's give an example which uses both steps, as well as some other things we have learned. We will show that $\exists x[\text{Large}(x) \wedge \text{LeftOf}(x, c)]$ follows from the following three premises.

1. $\forall x[\text{Cube}(x) \rightarrow \text{Large}(x)]$
2. $\forall x[\text{Large}(x) \rightarrow \text{LeftOf}(x, c)]$
3. $\text{Cube}(d)$

This is a rather obvious result, which is all the better for illustrating the obviousness of these steps.

Proof: Using universal instantiation, we get

$$\text{Cube}(d) \rightarrow \text{Large}(d)$$

and

$$\text{Large}(d) \rightarrow \text{LeftOf}(d, c)$$

Applying modus ponens to (3) and the first of these gives us $\text{Large}(d)$. Another application of modus ponens gives us $\text{LeftOf}(d, c)$. Now (using conjunction introduction) we have

$$\text{Large}(d) \wedge \text{LeftOf}(d, c).$$

Finally, applying existential introduction gives us our desired conclusion,

$$\exists x[\text{Large}(x) \wedge \text{LeftOf}(x, c)].$$

Before leaving this section, we should point out that there are ways to prove existential statements other than by existential generalization. In particular, to prove $\exists x P(x)$ we could use proof by contradiction, assuming $\neg\exists x P(x)$ and deriving a contradiction. This method of proceeding is somewhat less satisfying, since it does not actually tell you which object it is that satisfies the condition $P(x)$. Still, it does show that there is some such object, which is all that is claimed. This was, in fact, the method we used in proving that there are irrational numbers x and y such that x^y is rational.

The method of existential instantiation

Existential instantiation is one of the more interesting and subtle methods of proof. It allows you to prove results when you are given an existential statement. Suppose our domain of discourse consists of all children, and you are told that some boy is at home. If you want to use this fact in your reasoning, you are of course not entitled to infer that Max is at home. Neither are you allowed to infer that Danny is at home. In fact, there is no particular boy of whom you can safely conclude that he is at home. So how should we proceed? What we could do is give a temporary name to one of the boys who is at home, and refer to him using that name, as long as we are careful not to use a name already used in the premises or the desired conclusion.

This sort or reasoning is used in everyday life, when we know that someone (or something) satisfies a certain condition, but do not know who (or what) satisfies it. For example, when Scotland Yard found out there was a serial killer at large, they dubbed him "Jack the Ripper," and used this name in reasoning about him. No one thought that this meant they knew who the killer was; rather, they simply introduced the name to refer to whoever was doing the killing. Note that if the town tailor were already called Jack the Ripper, then the detectives' use of this name would (probably) have been a gross injustice.

This is a basic strategy used when giving proofs in FOL. If we have correctly proven that $\exists x S(x)$ then we can give a name, say c, to one of the objects satisfying $S(x)$, as long as the name is not one that is already in use. We may then assert $S(c)$ and use it in our proof. This is the rule known as "existential instantiation" or "existential elimination."

Generally, when existential instantiation is used in a mathematical proof, this will be marked by an explicit introduction of a new name. For example, the author of the proof might say, "So we have shown that there is a prime number between n and m. Call it p." Another phrase that serves the same function is: "Let p be such a prime number."

Let's give an example of how this rule might be used, by modifying our preceding example. The desired conclusion is the same, $\exists x(\mathsf{Large}(x) \land \mathsf{LeftOf}(x, c))$, but we will now take the premises to be:

1. $\forall x[\mathsf{Cube}(x) \rightarrow \mathsf{Large}(x)]$
2. $\forall x[\mathsf{Large}(x) \rightarrow \mathsf{LeftOf}(x, c)]$
3. $\exists x\ \mathsf{Cube}(x)$

The first two are just as before, but the third is weaker. But we can still obtain the same conclusion.

We would like to eliminate the $\exists$ in our third premise, since then we would be back to the case we have already examined. How then should we proceed? The proof would take the following form.

> **Proof:** We first note that the third premise assures us that there is at least one cube. Let e be one of these cubes. We can now proceed just as in our earlier reasoning. Applying the first premise, we see that e must be large. (What steps are we using here?) Applying the second premise, we see that e must also be left of c. Thus we have shown that e is both large and left of c. Our desired conclusion follows (by what inference step?) from this claim.

In applying existential instantiation, it is very important to make sure you use a new name, not one that is already in use. Looking at the above example shows why. Suppose we had thoughtlessly used the name c for the cube we named e. Then we would have been able to prove $\exists x\ \mathsf{LeftOf}(x, x)$, which is impossible. But our original premises are obviously satisfiable: they are true in many different worlds. (See Problem 22.) So if we do not observe this condition, we can be led from true premises to false (even impossible) conclusions.

The method of general conditional proof

One of the most important methods of proof involves reasoning about an arbitrary object of a particular kind, in order to prove a universal claim about all such objects. It is known as the method of *general* conditional proof. It is a more powerful version of conditional proof, similar in spirit to the method of existential instantiation just discussed.

Let's start out with an example. This time let us assume that the domain of discourse consists of students at a particular college. We suppose that we are given a bunch of information about these students in the form of premises. Finally, let us suppose we are able to prove from these premises that Sandy, a math major, is smart. Under what conditions would we be entitled to infer that every math major at the school is smart?

At first sight, it seems that we could never draw such a conclusion, unless there were only one math major at the school. After all, it does not follow from the fact that one math major is smart that all math majors are. But what if our proof that Sandy is smart uses *nothing at all* that is particular to Sandy? What if the proof would apply equally well to any math major? Then it seems that we should be able to conclude that every math major is smart.

How might one use this in a real example? Let us suppose that our premises are:

1. Anyone who passes Logic 101 with an A is smart.
2. Every math major has passed Logic 101 with an A.

Our desired conclusion is that every math major is smart. Our reasoning proceeds as follows.

> **Proof:** Let "Sandy" refer to any one of the math majors. By the second premise, Sandy passed Logic 101 with an A. By the first premise, then, Sandy is smart. But since Sandy is an arbitrarily chosen math major, it follows that every math major is smart.

This method of reasoning is used at every turn in doing mathematics. The general form is the following. Suppose we want to prove $\forall x[P(x) \rightarrow Q(x)]$ from some premises. The most straightforward way to proceed is to choose a name that is not in use, say c, assume P(c), and prove Q(c). If you are able to do this, then you are entitled to infer the desired result.

Let's look at another example. Suppose we wanted to prove that every prime number has an irrational square root. To apply general conditional proof, we begin by assuming that p is an arbitrary prime number. Our goal is to show that $\sqrt{p}$ is irrational. If we can do this, we will have established the general claim. We have already proven that this holds if $p = 2$. But our proof relied on specific facts about 2, and so the general claim certainly doesn't follow from our proof. The proof, however, can be generalized to show what we want. Here, for the record, is the generalization.

> **Proof:** Assume that p is any prime number. Since p is prime, it follows that if p divides a square, say k^2, then it divides k. Hence, if p divides k^2, p^2 also divides k^2. Now assume, for proof by contradiction, that $\sqrt{p}$ is rational. Write it in lowest terms as $\sqrt{p} = n/m$. In particular, we can make sure that p does not divide both n and m without remainder. Now, squaring both sides, we see that
>
> $$p = \frac{n^2}{m^2}$$
>
> and hence
>
> $$pm^2 = n^2$$
>
> But then it follows that p divides n^2, and so, as we have seen,

p divides n and p^2 divides n^2. But from the latter of these it follows that p^2 divides pm^2 so p divides m^2. But then p divides m. So we have shown that p divides both n and m, contradicting our choice of n and m. This contradiction shows that $\sqrt{p}$ is indeed irrational.

In formal general conditional proof systems of deduction, the method of general conditional proof is broken down into two parts, conditional proof and a method for proving completely general claims, claims of the form ∀xS(x). The latter method is called "universal generalization" or "universal introduction." It tells us that if we are able to introduce a new name c to stand for a completely arbitrary member of the domain of discourse and go on to prove the sentence S(c), then we can conclude ∀xS(x).

Here is a very simple example. Suppose we are given the premises

1. ∀x[R(x) → S(x)]
2. ∀xR(x)

and want to prove ∀xS(x).

Proof: We begin by taking a new name d, and think of it as standing for any member of the domain of discourse. Applying universal instantiation twice, once to each premise, gives us

1. R(d) → S(d)
2. R(d)

By modus ponens, we conclude S(d). But d denotes an arbitrary object in the domain, so our conclusion ∀xS(x) follows by universal generalization.

Any proof using general conditional proof could be converted into a proof using universal generalization, together with the method of conditional proof, in the following way. Suppose we have managed to prove ∀x[P(x) → Q(x)] using general conditional proof. Here is how we would go about proving it with universal generalization instead. First we would introduce a new name c, and think of it as standing for an arbitrary member of the domain of discourse. We know we can then prove P(c) → Q(c) using ordinary conditional proof, since that is what we did in our original proof. But then, since c stands for an arbitrary member of the domain, we can use universal generalization to get ∀x[P(x) → Q(x)].

This is how formal systems of deduction get by without having an explicit rule of general conditional proof. But really, we can equally well think of universal generalization as a special case of general conditional proof. After all, if we wanted to prove ∀xS(x) we could apply general conditional proof to the logically equivalent sentence ∀x[x = x → S(x)].

Or, if had a vacuous predicate like Thing(x), we could use general conditional proof to obtain ∀x[Thing(x) → S(x)].

We have chosen to emphasize general conditional proof since it is really the method used in giving rigorous informal proofs. The division of this method into conditional proof and universal generalization is a clever trick, but it does not correspond well to actual reasoning. This is at least in part due to the fact that universal noun phrases of English are always restricted by some common noun, if only the noun *thing*. The natural counterparts of such statements in FOL have the form ∀x[P(x) → Q(x)], which is why we typically prove them by general conditional proof. When we extend our formal system $\mathcal{F}$ to include quantifier rules, we will use universal generalization. However, we will introduce a formal counterpart of general conditional proof in the wider system $\mathcal{F}'$.

Remember

Let S(x), P(x), and Q(x) be wffs.

1. Existential Instantiation: If you have proven ∃xS(x) then you may choose a *new* constant symbol c and assume S(c).
2. General Conditional Proof: If you want to prove ∀x[P(x) → Q(x)] then you may choose a *new* constant symbol c, assume P(c), and prove Q(c).
3. Universal Generalization: If you want to prove ∀xS(x) then you may choose a *new* constant symbol c and prove S(c).

When we deal with sentences that contain both ∀ and ∃, we will see that the method of general conditional proof must be refined a bit. We will do just that in the next chapter.

Exercises and Problems

Problem$_\diamond$ 22. Build a world where the following are all true. Save it as World 5.22.

1. ∀x[Cube(x) → Large(x)]
2. ∀x[Large(x) → LeftOf(x, c)]
3. ∃x Cube(x)

Verify the claim made in the preceding discussion that if we do not observe the "newness condition" on existential instantiation, we can be led from true premises to a false conclusion.

Problem 23. Give a logical analysis of the following purported proof. Make explicit each valid step and method of proof used. If there are any invalid steps, identify them and indicate how you know they are invalid.

If there is a mistake, can you patch it up by giving a correct proof of the conclusion from the premises? The premises are:

1. $\forall x[(\mathsf{Brillig}(x) \lor \mathsf{Tove}(x)) \to (\mathsf{Mimsy}(x) \land \mathsf{Gyre}(x))]$
2. $\forall y[(\mathsf{Slithy}(y) \lor \mathsf{Mimsy}(y)) \to \mathsf{Tove}(y)]$
3. $\exists x\ \mathsf{Slithy}(x)$

The purported conclusion is $\exists x[\mathsf{Slithy}(x) \land \mathsf{Mimsy}(x)]$. Here is the purported proof:

> By the third premise, we know that something in the domain of discourse is slithy. Let b be one of these slithy things. By the second premise, we know that b is a tove. By the first premise, we see that b is mimsy. Thus, b is both slithy and mimsy. Hence something is both slithy and mimsy.

Problem 24. Give a logical analysis of the following purported proof. Make explicit each valid step and method of proof used. If there are any invalid steps, identify them and indicate how you know they are invalid. If there is a mistake, can you patch it up by giving a correct proof of the conclusion from the premises? The premises are:

1. $\forall x[\mathsf{Brillig}(x) \to (\mathsf{Mimsy}(x) \land \mathsf{Slithy}(x))]$
2. $\forall y[(\mathsf{Slithy}(y) \lor \mathsf{Mimsy}(y)) \to \mathsf{Tove}(y)]$
3. $\forall x[\mathsf{Tove}(x) \to (\mathsf{Outgrabe}(x, b) \land \mathsf{Brillig}(x))]$

The purported conclusion is $\forall z[\mathsf{Brillig}(z) \leftrightarrow \mathsf{Mimsy}(z)]$. Here is the purported proof:

> In order to prove the conclusion, it suffices to prove the logically equivalent sentence obtained by conjoining the following two sentences:
>
> $\forall x[\mathsf{Brillig}(x) \to \mathsf{Mimsy}(x)]$
> $\forall x[\mathsf{Mimsy}(x) \to \mathsf{Brillig}(x)]$
>
> We prove these by the method of general conditional proof, in turn. First, let b be anything that is Brillig. Then by (1) it is both Mimsy and Slithy. Hence it is Mimsy, as desired. Thus we have the first of our two desired results.
> To prove the second, let b be anything that is Mimsy. By the second premise, b is also Tove. But then by the final premise, b is Brillig, as desired. This concludes the proof.

Problem 25. Give a logical analysis of the following purported proof. Make explicit each valid step and method of proof used. If there are any invalid steps, identify them and indicate how you know they are invalid. If there is a mistake, can you patch it up by giving a correct proof of the conclusion from the premises? The premises are:

1. ∀x[(Brillig(x) ∧ Tove(x)) → Mimsy(x)]
2. ∀y[(Tove(y) ∨ Mimsy(y)) → Slithy(y)]
3. ∃x Brillig(x) ∧ ∃x Tove(x)

The purported conclusion is ∃z Slithy(z) Here is the purported proof:

> By the third premise, we know that there are brillig toves. Let b be one of them. By the first premise, we know that b is mimsy. By the second premise, we know that b is slithy. Hence there is something that is slithy.

Problem◇ 26. Assume the following premises:

1. ∀y[Cube(y) ∨ Dodec(y)]
2. ∀x[Cube(x) → Large(x)]
3. ∃x¬Large(x)

Does it follow that ∃x Dodec(x)? If so, give a proof. If not, build a world in which the premises are true and the conclusion false. Save it as World 5.26.

Problem◇ 27. Assume the same premises as in Problem 26. Does it follow that ∃x[Dodec(x) ∧ Small(x)]? If so, give a proof. If not, build a world in which the premises are true but the conclusion false. Save it as World 5.27.

Problem◇ 28. Assume the following premises:

1. ∀x[Cube(x) ∨ Dodec(x)]
2. ∀x[Cube(x) → (Large(x) ∧ LeftOf(c, x))]
3. ∀x[¬Small(x) → Tet(x)]

Does it follow that Small(c)? If so, give a proof. If not, build a world in which the premises are true but the conclusion false. Save it as World 5.28.

Problem◇ 29. Assume the same premises as in Problem 28. Does it follow that Dodec(c)? If so, give a proof. If not, build a world in which the premises are true but the conclusion false. Save it as World 5.29.

Problem◇ 30. Assume the following premises:

1. ∀x[Cube(x) ∨ (Tet(x) ∧ Small(x))]
2. ∃x[Large(x) ∧ BackOf(x, c)]

Does it follow that ∃x[FrontOf(c, x) ∧ Cube(x)]? If so, give a proof. If not, build a world in which the premises are true but the conclusion false. Save it as World 5.30.

Problem◇ 31. Assume the same premises as in Problem 30. Does it follow that ¬∃x[Small(x) ∧ BackOf(x, c)]? If so, give a proof. If not, build a world in which the premises are true but the conclusion false. Save it as World 5.31.

Problem 32. Translate the following premises and purported conclusion into FOL and determine whether or not the conclusion follows from the premises. If it does, give a proof.

1. Every child is either right-handed or intelligent.
2. No intelligent child eats liver.
3. There is a child who eats liver and onions.
4. Conclusion: There is a right-handed child who eats onions.

Problem 33. Are sentences (1) and (2) in Problem 16 logically equivalent? If so, give a proof. If not, explain why not.

Problem 34. Let us work in the first-order language of arithmetic with the added predicates Even(x), Prime(x), and DivisibleBy(x, y), where these have the obvious meanings (the last means that the natural number y divides the number x without remainder.) Prove the following results:

1. $\exists y[\mathsf{Prime}(y) \land \mathsf{Even}(y)]$
2. $\forall x[\mathsf{Even}(x) \leftrightarrow \mathsf{Even}(x^2)]$
3. $\forall x[\mathsf{DivisibleBy}(x^2, 3) \rightarrow \mathsf{DivisibleBy}(x^2, 9)]$

In some cases, you have already done all the hard work in earlier problems.

5.11 Formal proofs

In this section we introduce the formal counterparts of the methods of proof involving the quantifiers. Again, we separate them into introduction and elimination rules.

Universal quantifier rules

The valid inference step of universal instantiation or elimination is easily formalized. Here is the schematic version of the rule:

Universal Elimination (∀ Elim):

| ∀xS(x)
| ⋮
▷ | S(c)

Here x stands for any variable, c stands for any individual constant (whether or not it has been used elsewhere in the proof), and S(c) stands for the result of replacing free occurrences of x in S(x) with c.

Next, let us formalize the more interesting method of universal generalization or introduction. This requires us to decide on a way to represent the fact that a constant symbol has been introduced to stand for an arbitrary object. We indicate this by means of a subproof with no assumptions, insisting that the constant in question occur only within that subproof. This will guarantee, for example, that the constant does not appear in the premises of the overall proof.

To remind ourselves of this crucial restriction, we will introduce a new graphical device, boxing the constant symbol in question and putting it at the top of the subproof. We will think of this as the formal analog of the English phrase "Let c denote an arbitrary object."

Universal Introduction (∀ Intro):

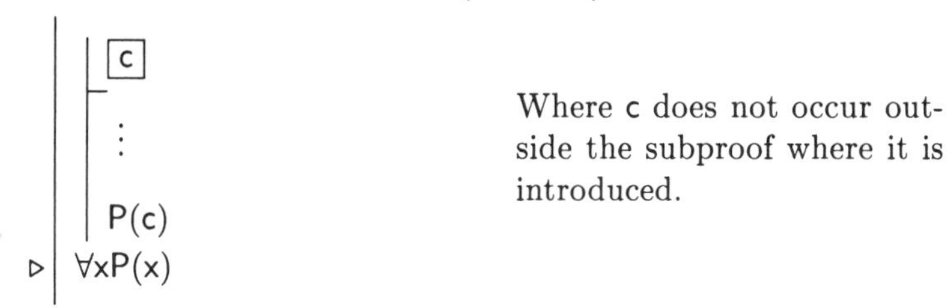

The requirement that c not occur outside the subproof in which it is introduced does not preclude it occurring within subproofs of that subproof. A sentence in a subproof of a subproof still counts as a sentence of the larger subproof.

When we give the justification for universal introduction, we will cite the subproof, as we do in the case of conditional introduction. However, rather than citing the subproof by referring to the first and last steps, we will refer to the introduction of the constant symbol, along with the last step.

Let us illustrate how to use these rules by giving a formal proof mirroring the informal proof given on page 140. We prove that ∀xS(x) follows from the premises ∀x[R(x) → S(x)] and ∀xR(x).

```
| 1. ∀x[R(x) → S(x)]
| 2. ∀xR(x)
|---
|  | [d]
|  |---
|  | 3. R(d) → S(d)          ∀ Elim: 1
|  | 4. R(d)                 ∀ Elim: 2
|  | 5. S(d)                 → Elim: 3, 4
| 6. ∀xS(x)                  ∀ Intro: [d]-5
```

Notice that the constant symbol d does not appear outside the subproof. It is newly introduced at the beginning of that subproof, and occurs nowhere else outside it. That is what allows the introduction of the universal quantifier in step 6.

Existential quantifier rules

Recall that in our discussion of informal proofs, existential introduction was a simple proof step, whereas the elimination of $\exists$ was a subtle method of proof. Thus, in presenting our formal system, we begin with the introduction rule.

Existential Introduction ($\exists$ Intro):

$$\begin{array}{l|l} & S(c) \\ & \vdots \\ \triangleright & \exists x S(x) \end{array}$$

Here too x stands for any variable, c stands for any individual constant, and $S(c)$ stands for the result of replacing free occurrences of x in $S(x)$ with c. Note that there may be other occurrences of c in $S(x)$ as well.

When we turn to the rule of existential elimination, we employ the same device as with universal introduction. If we have proven $\exists x S(x)$, then we introduce a new constant symbol, say c, along with the assumption that the object denoted by c satisfies the formula $S(x)$. If, from this assumption, we can derive some sentence Q not using the constant c, then we can conclude that Q follows from the original premises.

Existential Elimination ($\exists$ Elim):

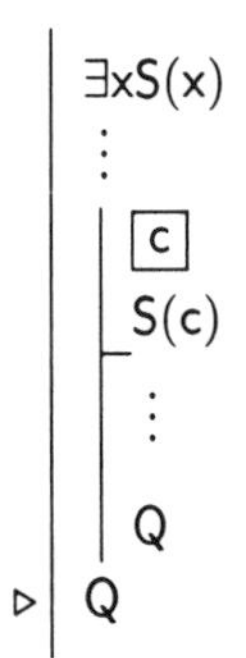

Where c does not occur outside the subproof where it is introduced.

We can think of the notation at the beginning of the subproof as the formal counterpart of the English "Let c be an arbitrary individual such that $S(c)$."

The rule of existential elimination is quite analogous to the rule of disjunction elimination, both formally and intuitively. With disjunction elimination, we have a disjunction and break into cases, one for each disjunct, and establish the same result in each case. With existential elimination, we can think of having one case for each object in the domain of discourse. We are required to show that, whichever object it is that satisfies the condition $\mathsf{S}(x)$, the same result Q can be obtained. If we can do this, we may conclude Q.

To illustrate the two existential rules, we will give a formal counterpart to the proof given on page 138.

1. $\forall x[\mathsf{Cube}(x) \to \mathsf{Large}(x)]$	
2. $\forall x[\mathsf{Large}(x) \to \mathsf{LeftOf}(x, c)]$	
3. $\exists x\, \mathsf{Cube}(x)$	
[e]	
4. $\mathsf{Cube}(e)$	
5. $\mathsf{Cube}(e) \to \mathsf{Large}(e)$	$\forall$ Elim: 1
6. $\mathsf{Large}(e)$	$\to$ Elim: 5, 4
7. $\mathsf{Large}(e) \to \mathsf{LeftOf}(e, c)$	$\forall$ Elim: 2
8. $\mathsf{LeftOf}(e, c)$	$\to$ Elim: 7, 6
9. $\mathsf{Large}(e) \wedge \mathsf{LeftOf}(e, c)$	$\wedge$ Intro: 6, 8
10. $\exists x(\mathsf{Large}(x) \wedge \mathsf{LeftOf}(x, c))$	$\exists$ Intro: 9
11. $\exists x(\mathsf{Large}(x) \wedge \mathsf{LeftOf}(x, c))$	$\exists$ Elim: 3, [e]-10

A worked example

Let's work through a moderately difficult proof, step by step. Suppose we want to give a formal proof that $\exists x \neg \mathsf{P}(x)$ follows from $\neg \forall x \mathsf{P}(x)$. We begin as usual with the skeleton of our proof:

1. $\neg \forall x \mathsf{P}(x)$
⋮
$\exists x \neg \mathsf{P}(x)$

Rather than start with an informal proof, we will construct an informal proof as we go. Since we are trying to prove an existential sentence, our first thought would be to use existential introduction, say by proving $\neg \mathsf{P}(c)$ for some c. In other words, we hope to fill in a proof of the following form:

1. $\neg \forall x \mathsf{P}(x)$	
⋮	
$\neg \mathsf{P}(c)$	
$\exists x \neg \mathsf{P}(x)$	$\exists$ Intro: ?

But if we think a bit about what our premise means, we see there is no hope of proving of any particular thing that it satisfies $\neg \mathsf{P}(x)$. From the fact that not everything satisfies $\mathsf{P}(x)$, we aren't going to prove of some specific c that $\neg \mathsf{P}(c)$. So this is surely a dead end.

This leaves only one possible route to our desired conclusion: proof by contradiction. Thus we will try to fill in the following:

```
| 1. ¬∀xP(x)
|---
|   | 2. ¬∃x¬P(x)
|   |---
|   |   ⋮
|   | Some contradiction
| ¬¬∃x¬P(x)                        ¬ Intro: 2–?
| ∃x¬P(x)                          ¬ Elim: ?
```

What contradiction can we hope to obtain here? Since our only premise is $\neg\forall x P(x)$, the most promising line of attack would be to try for a proof of $\forall x P(x)$ using universal generalization. Let's try this strategy:

```
| 1. ¬∀xP(x)
|---
|   | 2. ¬∃x¬P(x)
|   |---
|   |   | [c]
|   |   |---
|   |   |   ⋮
|   |   | P(c)
|   | ∀xP(x)                       ∀ Intro: [c]-?
|   | ∀xP(x) ∧ ¬∀xP(x)             ∧ Intro: ?, 1
| ¬¬∃x¬P(x)                        ¬ Intro: 2–?
| ∃x¬P(x)                          ¬ Elim: ?
```

So now the problem is how to prove $P(c)$. Let's think about how we would prove this informally. If $P(c)$ were not the case, then we would have $\neg P(c)$, and hence $\exists x \neg P(x)$. But this contradicts the assumption at step 2. This shows us how to fill in the rest of the proof.

```
| 1. ¬∀xP(x)
|---
|   | 2. ¬∃x¬P(x)
|   |---
|   |   | [c]
|   |   |---
|   |   |   | 3. ¬P(c)
|   |   |   |---
|   |   |   | 4. ∃x¬P(x)               ∃ Intro: 3
|   |   |   | 5. ∃x¬P(x) ∧ ¬∃x¬P(x)    ∧ Intro: 4, 2
|   |   | 6. ¬¬P(c)                    ¬ Intro: 3–5
|   |   | 7. P(c)                      ¬ Elim: 6
|   | 8. ∀xP(x)                        ∀ Intro: [c]-7
|   | 9. ∀xP(x) ∧ ¬∀xP(x)              ∧ Intro: 8, 1
| 10. ¬¬∃x¬P(x)                        ¬ Intro: 2–9
| 11. ∃x¬P(x)                          ¬ Elim: 10
```

This completes our formal proof of $\exists x \neg P(x)$ from the premise $\neg \forall x P(x)$.

Exercises and Problems

Problem 35. Write out an informal proof that corresponds to the completed formal proof given in the previous section.

Problem 36. Recall from Problem 23 on page 141 that you were asked to give an analysis of an informal argument, one that purported to show that $\exists x[\mathsf{Slithy}(x) \land \mathsf{Mimsy}(x)]$ is a consequence of the following premises.

1. $\forall x[(\mathsf{Brillig}(x) \lor \mathsf{Tove}(x)) \rightarrow (\mathsf{Mimsy}(x) \land \mathsf{Gyre}(x))]$
2. $\forall y[(\mathsf{Slithy}(y) \lor \mathsf{Mimsy}(y)) \rightarrow \mathsf{Tove}(y)]$
3. $\exists x\ \mathsf{Slithy}(x)$

Give a formal counterpart to this argument and assess its validity. If the argument was valid, your formal proof should be too. If the informal argument was invalid, show where the formal proof is wrong.

Problem 37. In Problem 25 on page 142 you were asked to give a logical analysis of an informal argument purporting to show that $\exists z\ \mathsf{Slithy}(z)$ follows from the following premises.

1. $\forall x[(\mathsf{Brillig}(x) \land \mathsf{Tove}(x)) \rightarrow \mathsf{Mimsy}(x)]$
2. $\forall y[(\mathsf{Tove}(y) \lor \mathsf{Mimsy}(y)) \rightarrow \mathsf{Slithy}(y)]$
3. $\exists x\ \mathsf{Brillig}(x) \land \exists x\ \mathsf{Tove}(x)$

Formalize the argument given and give an analysis of it. If it is invalid, can you give a valid proof of the same conclusion?

Problem 38. Give a formal proof of $\exists x\ \mathsf{Dodec}(x)$ from the following premises:

1. $\forall y[\mathsf{Cube}(y) \lor \mathsf{Dodec}(y)]$
2. $\forall x[\mathsf{Cube}(x) \rightarrow \mathsf{Large}(x)]$
3. $\exists x \neg \mathsf{Large}(x)$

Problem$_\diamond$ 39. Some of the following are valid, some are not. In each, either give a formal proof of the inference or construct a world in which the premise is true but the conclusion false. Save the world for part x as World 5.39.x.

1. $\exists x\ \mathsf{Cube}(x) \land \exists x\ \mathsf{Small}(x)$ from the premise $\exists x(\mathsf{Cube}(x) \land \mathsf{Small}(x))$.
2. $\exists x(\mathsf{Cube}(x) \land \mathsf{Small}(x))$ from the premise $\exists x\ \mathsf{Cube}(x) \land \exists x\ \mathsf{Small}(x)$.
3. $\exists x(\mathsf{Cube}(x) \land \mathsf{Small}(d))$ from the premise $\exists x\ \mathsf{Cube}(x) \land \mathsf{Small}(d)$.
4. $\forall x\ \mathsf{Cube}(x) \lor \forall x\ \mathsf{Small}(x)$ from the premise $\forall x(\mathsf{Cube}(x) \lor \mathsf{Small}(x))$.
5. $\forall x(\mathsf{Cube}(x) \lor \mathsf{Small}(x))$ from the premise $\forall x\ \mathsf{Cube}(x) \lor \forall x\ \mathsf{Small}(x)$.

Problem 40. (Change of bound variables) Give formal proofs of the following:

1. $\forall x P(x)$ from the premise $\forall y P(y)$.
2. $\exists x P(x)$ from the premise $\exists y P(y)$.

Problem 41. (Quantifiers and negation) Give formal proofs of the following:

1. $\neg\forall x P(x)$ from the premise $\exists x \neg P(x)$.
2. $\neg\exists x P(x)$ from the premise $\forall x \neg P(x)$.
3. $\forall x \neg P(x)$ from the premise $\neg\exists x P(x)$.

5.12 Function symbols, revisited

In Section 2.5, we discussed first-order languages with function symbols. There we said that function symbols allow us to form "complex names" out of other names. For example, father(father(Max)) refers to Max's father's father. And $(1 + (1 + 1))$ refers to the number 3.

Now that we have variables at our disposal, we allow them to occur inside the terms that are constructed by means of function symbols. Thus, for example, we could have a term father(father(x)), which would be useful in expressing things like

$$\forall x\ \mathsf{NicerThan}(\mathsf{father}(\mathsf{father}(x)), \mathsf{father}(x))$$

which says that everyone's paternal grandfather is nicer than their father, a false belief held by many children.

Thus, when there are function symbols in the language, terms containing variables can be used inside atomic wffs. So, for example, the following would be atomic wffs:

$$\mathsf{TallerThan}(\mathsf{father}(x), x)$$
$$\mathsf{TallerThan}(\mathsf{mother}(x), \mathsf{father}(y))$$
$$(x + z) < (x + y)$$

And then, of course, these can occur inside more complex wffs, including sentences. We have, in fact, been using function symbols in some of our examples and problems, without comment. Thus, for example, in the sentence

$$\forall y[\mathsf{Even}(y) \leftrightarrow \mathsf{Even}(y^2)]$$

what we really should have written was the following, which has an explicit use of the function symbol $\times$:

$$\forall y[\mathsf{Even}(y) \leftrightarrow \mathsf{Even}(y \times y)]$$

Function symbols are extremely useful and important in applications of FOL. We close this chapter with some problems that use function symbols.

Exercises and Problems

Problem 42. Consider the first-order language with function symbols mother and father, plus a variety of new names. Let us imagine that the following table lists some fathers and mothers of various people:

Child	Mother	Father
Claire	Mary	Jon
Melanie	Mary	Jon
Mary	Helen	Jim
Helen	Anna	William
Jon	Evelyn	Kenneth
Jim	Addie	Archie
Evelyn	Ellen	John

Here are some atomic wffs, each with a single free variable x. For each, pick a person for x that satisfies the wff, if you can. If there is no such person, say so.

1. mother(x) = Ellen
2. father(x) = Jon
3. mother(father(x)) = Mary
4. father(mother(x)) = John
5. mother(father(x)) = Addie
6. father(mother(father(x))) = John
7. father(father(mother(x))) = Archie
8. father(father(Jim)) = x
9. father(father(mother(Claire))) = x

Problem 43. Using the table from Problem 42, plus the following height table, figure out which of the sentences listed below are true. (Assume that the domain of discourse consists of the people listed in the table.)

Person	height	Person	height
Claire	4′2″	Melanie	5′2″
Addie	5′3″	Ellen	5′4″
Anna	5′7″	William	5′8″
Mary	5′6″	Helen	5′6″
Evelyn	5′7″	Archie	5′8″
Kenneth	5′10″	Jim	6′0″
Jon	6′5″	John	5′11″

1. ∃xTallerThan(x, mother(x))
2. ∃yTallerThan(mother(mother(y)), mother(father(y)))

3. $\forall x[\mathsf{height}(x) < 5' \rightarrow \mathsf{TallerThan}(\mathsf{father}(x), \mathsf{Jim})]$
4. $\forall z[z \neq \mathsf{father}(\mathsf{Claire}) \rightarrow \mathsf{TallerThan}(\mathsf{father}(\mathsf{Claire}), z)]$

Problem 44. Assume we are working in an extension of the first-order language of arithmetic with the additional predicates Even(x) and Prime(x). Express the following in this language using the function symbol $\times$ explicitly, as in $z \times z$, rather than z^2.

1. No square is prime.
2. Some square is odd.
3. The square of any number greater than 1 is greater than the number itself.

Which of these sentences are true? (The number 1 does not count as a prime.)

5.13 Alternative notation

The notation we have been using for the quantifiers is currently the most popular. An older notation that is still in some use employs (x) for $\forall x$. Thus, for example, in this notation our

$$\forall x[\mathsf{Tet}(x) \rightarrow \mathsf{Small}(x)]$$

would be written

$$(x)\ [\mathsf{Tet}(x) \rightarrow \mathsf{Small}(x)]$$

Another notation which is occasionally used exploits the similarity between universal quantification and conjunction by writing $\bigwedge x$ instead of $\forall x$. In this notation our sentence would be rendered

$$\bigwedge x\ [\mathsf{Tet}(x) \rightarrow \mathsf{Small}(x)]$$

Finally, you will sometimes encounter the universal quantifier written Πx, as in

$$\Pi x[\mathsf{Tet}(x) \rightarrow \mathsf{Small}(x)]$$

Similar variants of $\exists x$ are in use. One version writes $(\exists x)$ or $(\mathsf{E}x)$. Other versions write $\bigvee x$ or Σx. Thus the following are notational variants of one another.

$$\exists x[\mathsf{Cube}(x) \wedge \mathsf{Large}(x)]$$
$$(\mathsf{E}x)[\mathsf{Cube}(x) \wedge \mathsf{Large}(x)]$$
$$\bigvee x\ [\mathsf{Cube}(x) \wedge \mathsf{Large}(x)]$$
$$\Sigma x[\mathsf{Cube}(x) \wedge \mathsf{Large}(x)]$$

Exercises and Problems

Problem 45. (Overcoming dialect differences) The following are all sentences of FOL. But they're in different dialects. Translate them into our dialect.

1. $\sim (x)(P(x) \supset Q(x))$
2. $\Sigma y((P(y) \equiv \overline{Q(y)}) \,\&\, R(y))$
3. $\overline{\bigwedge x\, P(x)} \equiv \bigvee x\, \overline{P(x)}$

Remember

The following table summarizes the alternative notations.

Our notation	Common equivalents
$\neg P$	$\sim P$, $\overline{P}$, $!P$, Np
$P \wedge Q$	$P\&Q$, $P\&\&Q$, $P \cdot Q$, PQ, Kpq
$P \vee Q$	$P \mid Q$, $P \parallel Q$, Apq
$P \rightarrow Q$	$P \supset Q$, Cpq
$P \leftrightarrow Q$	$P \equiv Q$, Epq
$\forall x S(x)$	$(x)S(x)$, $\bigwedge x\, S(x)$, $\Pi x S(x)$
$\exists x S(x)$	$(\exists x)S(x)$, $(Ex)S(x)$, $\bigvee x\, S(x)$, $\Sigma x S(x)$

6

Sentences with More than One Quantifier

The previous chapter presented the basic theory of quantifiers, but our problems and exercises were restricted to sentences that contain only a single quantifier. In practice, however, this is far too restricted. You need to be fluent enough not to be thrown for a loop by sentences that have lots of quantifiers.

6.1 Multiple uses of a single quantifier symbol

In this section we study sentences that have multiple instances of $\forall$ or multiple instances of $\exists$, but no mixing of the two. Here are a couple of simple sentences that contain multiple quantifiers:

$$\exists y \exists z[\mathsf{Cube}(y) \land \mathsf{Tet}(z) \land \mathsf{LeftOf}(y, z)]$$
$$\forall x \forall y[(\mathsf{Cube}(x) \land \mathsf{Tet}(y)) \rightarrow \mathsf{LeftOf}(x, y)]$$

Try to guess what these say. You should have no trouble. The first of them says that some cube is left of some tetrahedron. The second says that every cube is left of every tetrahedron.

In these examples, all the quantifiers are out in front (in what we will come to call "prenex form") but there is no need for them to be. The same claims could be expressed by the following sentences:

$$\exists y[\mathsf{Cube}(y) \land \exists z(\mathsf{Tet}(z) \land \mathsf{LeftOf}(y, z))]$$
$$\forall x[\mathsf{Cube}(x) \rightarrow \forall y(\mathsf{Tet}(y) \rightarrow \mathsf{LeftOf}(x, y))]$$

It is easy to see that these make the same claims as the first pair, even though, in the case of the universal claim, the structure of the FOL sentence has changed considerably. Later we will introduce principles that will make it clear why they are equivalent sentences.

There is one slightly tricky point that arises with the use of multiple quantifiers. It is exhibited by the following sentence:

$$\forall x \forall y[(\mathsf{Cube}(x) \land \mathsf{Cube}(y)) \rightarrow (\mathsf{LeftOf}(x, y) \lor \mathsf{RightOf}(x, y))]$$

Imagine that we are evaluating this sentence in a world with five cubes lined up in the front row. Do you think the sentence would be true in this world? (You might want to try it.)

One natural way to read this sentence is as claiming that if x and y are cubes, then either x is left of y or x is right of y. But there is a conversational implicature in this way of speaking, one that is very misleading. It suggests that x and y are distinct cubes, but this is not part of the claim made by the first-order sentence. In fact, our sentence is false in the imagined world, as it must be in *any* world that contains at least one cube. For if it were true, then so would be the following instance of it (using b to name our cube):

$$\forall y[(\mathsf{Cube}(b) \land \mathsf{Cube}(y)) \rightarrow (\mathsf{LeftOf}(b,y) \lor \mathsf{RightOf}(b,y))]$$

This in turn is a universal claim, and so must also be true of the object b. But when we substitute b in for y we get

$$(\mathsf{Cube}(b) \land \mathsf{Cube}(b)) \rightarrow (\mathsf{LeftOf}(b,b) \lor \mathsf{RightOf}(b,b))]$$

which is obviously false, since b is neither left of nor right of itself.

If we really wanted to express the claim that every cube is to the left or right of every *other* cube, then we could write:

$$\forall x \forall y[(\mathsf{Cube}(x) \land \mathsf{Cube}(y) \land x \neq y) \rightarrow (\mathsf{LeftOf}(x,y) \lor \mathsf{RightOf}(x,y))]$$

This, of course, is true in the world we described.

Exercises and Problems

Exercise$_\diamond$ 1. (Evaluating multiple quantifier sentences) Open up Peano's World and Peano's Sentences, and evaluate the sentences in the world. If you have trouble with any, play the game (several times if necessary) until you see where you are going wrong.

Problem$_\diamond$ 2. (Building a world) Open Ramsey's Sentences. Build a world in which sentences 1–10 are all true at once. These sentences all make either "particular" claims (that is, they contain no quantifiers) or "existential" claims (that is, they assert that things of a certain sort exist). Consequently, you could make them true by successively adding objects to the world. But part of the exercise is to make them all true *with as few objects as possible.* You should be able to do it with a total of six objects. So rather than adding objects for each new sentence, try adjusting the world, and only add new objects when necessary. Again, be sure to go back and check that all the sentences are true when you are finished. Save the world as World 6.2. [Hint: To make all the sentences true with this small a world, one of the objects will have to have two names.]

Problem$_\diamond$ 3. (Modifying the world) Sentences 11-20 of Ramsey's Sentences all make "universal" claims. That is, they all say that every object in the world has some property or other. Check to see whether the world you have built in Problem 2 satisfies the universal claims expressed by these sentences. If not, modify the world so it makes all 20 sentences true at once. Once you have built this world, save it as World 6.3.

Problem$_\diamond$ 4. (Expanding a world) In the real world, things change in various ways. They come, move around, and go. And as things change, so do the truth values of sentences.

- In this exercise, the goal is to change World 6.3 to make as many of Ramsey's sentences false as you can. But here's the catch: you can only add objects of various sizes and shapes; don't change the existing objects in any way. Save the modified world as World 6.4.
- (⋆) Do you notice anything about which sentences you can make false in this way and which you cannot? Try to give a fairly clear and intuitive account of which sentences you cannot make false in this way. We will return to this topic in a later problem.

Exercise$_\diamond$ 5. (Evaluating multiple quantifier sentences) This exercise deals with multiple quantifiers and their interaction with negation. Open Frege's Sentences. Work through them one at a time, assessing their truth in Peirce's World. The game should come in very handy here.

Problem$_\diamond$ 6. Consider the following premises:

1. $\forall x \forall y[\mathsf{LeftOf}(x, y) \rightarrow \mathsf{Larger}(x, y)]$
2. $\forall x[\mathsf{Cube}(x) \rightarrow \mathsf{Small}(x)]$
3. $\forall x[\mathsf{Tet}(x) \rightarrow \mathsf{Large}(x)]$
4. $\forall x \forall y[(\mathsf{Small}(x) \wedge \mathsf{Small}(y)) \rightarrow \neg \mathsf{Larger}(x, y)]$

Is the following sentence a logical consequence of these premises?

$$\neg \exists x \exists y[\mathsf{Cube}(x) \wedge \mathsf{Cube}(y) \wedge \mathsf{RightOf}(x, y)]$$

If so, give a proof. If not, build a world where all the premises are true but the conclusion is false. Save it as World 6.6.

Problem$_\diamond$ 7. Use the same premises as in Problem 6. Is the following sentence a consequence of these premises?

$$\forall z[\mathsf{Medium}(z) \rightarrow \mathsf{Tet}(z)]$$

If so, give a proof. If not, build a world where the premises are all true but where the conclusion is false. Save it as World 6.7.

Problem$_\diamond$ 8. Use the same premises as in Problem 6. Is the following sentence a consequence of these premises?

$$\forall z \forall w[(\mathsf{Tet}(z) \wedge \mathsf{Cube}(w)) \rightarrow \mathsf{LeftOf}(z, w)]$$

If so, give a proof. If not, build a world where the premises are all true but where the conclusion is false. Save it as World 6.8.

Problem 9. Give a logical analysis of the following purported proof. The premises are:

1. $\forall x \forall y \forall z[(\text{Outgrabe}(x, y) \wedge \text{Outgrabe}(y, z)) \rightarrow \text{Outgrabe}(x, z)]$
2. $\forall x \forall y[\text{Outgrabe}(x, y) \rightarrow \text{Outgrabe}(y, x)]$
3. $\exists x \exists y\ \text{Outgrabe}(x, y)$

The purported conclusion is $\forall x\ \text{Outgrabe}(x, x)$. The proof is as follows:

> Applying existential instantiation to the third premise, let b and c be arbitrary objects in the domain of discourse such that Outgrabe(b,c). By the second premise, we also have Outgrabe(c,b). Applying the first premise (with x = z = b and y = c) we get Outgrabe(b,b). But b was arbitrary. Thus by universal generalization, $\forall x\ \text{Outgrabe}(x, x)$.

6.2 Mixed quantifiers

Now that you have plenty of experience with quantifiers, we will discuss the important case in which universal and existential quantifiers get mixed together.

Consider the following sentence:

$$\forall x[\text{Cube}(x) \rightarrow \exists y(\text{Tet}(y) \wedge \text{LeftOf}(x, y))]$$

This has the general form $\forall x[P(x) \rightarrow Q(x)]$ that we have seen many times before. It says that every cube has some property or other. What property? The property expressed by $\exists y(\text{Tet}(y) \wedge \text{LeftOf}(x, y))$, that is, the property of being left of a tetrahedron. Thus our original sentence claims that every cube is to the left of a tetrahedron.

This could also be expressed in a number of other ways. Here is the most important alternative:

$$\forall x \exists y[\text{Cube}(x) \rightarrow (\text{Tet}(y) \wedge \text{LeftOf}(x, y))]$$

We will see that these are equivalent later. The first version is more natural as a translation of the English, *Every cube is left of some tetrahedron,* but the second has the advantage of being in so-called prenex form, a form that we will discuss in due course.

When we have a sentence with a string of mixed quantifiers, as in the last sentence, the order of the quantifiers is very important. This is something we have not had to worry about before, with sentences that contain only universal or only existential quantifiers. Clearly, the sentence $\forall x \forall y\ \text{Likes}(x, y)$ is logically equivalent to the sentence where the order of the quantifiers is reversed: $\forall y \forall x\ \text{Likes}(x, y)$. They are both true just in case everything in the domain of discourse likes everything in the domain of discourse. Similarly, $\exists x \exists y\ \text{Likes}(x, y)$ is logically equivalent to $\exists y \exists x\ \text{Likes}(x, y)$. However, this is not the case when the quantifiers are mixed. $\forall x \exists y$ Likes(x, y) says that everyone likes someone, whereas $\exists y \forall x$ Likes(x, y) says that there is someone whom everyone likes. This is a far stronger claim. So when dealing with mixed quantifiers, you must be very sensitive to the order of quantifiers. We will learn more about getting the order of quantifiers right in the sections that follow.

Remember

When you are dealing with mixed quantifiers, the order is very important. $\forall x \exists y$R(x, y) does not logically imply $\exists y \forall x$R(x, y).

Exercises and Problems

Exercise$_\diamond$ 10. (Evaluating mixed quantifier sentences with identity) Open Leibniz's World and use it to evaluate the sentences in Leibniz's Sentences. If your assessment of a sentence is wrong, use the game to figure out why.

Problem$_\diamond$ 11. (Building a world) Open Buridan's Sentences. Build a world in which all ten sentences are true at once. Save the world, calling it World 6.11.

Problem$_\diamond$ 12. (Consequence) The following English sentences are all consequences of the ten sentences in Buridan's Sentences.

1. *There are no cubes.*
2. *Some tetrahedron is not large.*
3. *Nothing is in back of* ***a***.
4. *Only large things are in back of* ***b***.

Because of this, they must be true in any world in which Buridan's sentences are all true. So of course they must be true in World 6.11, no matter how you built it.

- Translate the sentences given above, adding them to the list in Buridan's Sentences. Verify that they are all true in World 6.11.
- For each of your translations in turn, modify World 6.11 so as to make the translation false. Save the world as World 6.12.x, where x is the number from the above list of the sentence which is false in this world. (So x is 1, 2, 3, or 4.) Then go back and see which of the original ten sentences have become false. One or more of them must have, since the translations are consequences of the original sentences.
- Show that sentence 9 is not a consequence of the remaining sentences in the file. [You can do that by changing the world so that sentence 9 is false while all the others are true. If sentence 9 *were* a consequence of the others, you wouldn't be able to do that.] Save your world as World 6.12.5.

- (★) Show that the following sentence (which says that there are at least two medium tetrahedra) is independent of those in Buridan's Sentences, i.e., neither it nor its negation is a consequence of those sentences.

$$\exists x\, \exists y\, (x \neq y \land \mathsf{Tet}(x) \land \mathsf{Tet}(y) \land \mathsf{Medium}(x) \land \mathsf{Medium}(y))$$

  You will do this by building two worlds, one in which this sentence is false and one in which its negation is false—but both of which make all of Buridan's sentences true. Save these two worlds as World 6.12.6 and World 6.12.7, respectively.

Exercise$_\diamond$ 13. (Evaluating more mixed quantifier sentences) Open Hilbert's Sentences and evaluate them in Peano's World.

Problem$_\diamond$ 14. (Building a world) Create a world *containing at most three objects* in which the ten sentences in Ockham's Sentences are all true. Save this world as World 6.14.

Problem$_\diamond$ 15. (Building a world) Create a world in which all ten sentences in Arnault's Sentences are true. Save the world as World 6.15.

Problem$_\diamond$ 16. (Describing a world) Open Peano's World. Start a new sentence file where you will describe some features of this world. Again, be sure to check each of your sentences to see that it is indeed a sentence and is true.

1. Notice that every dodecahedron is small. Use your first sentence to say this.
2. State the fact that there is a medium sized cube.
3. Next, assert that there are at least two cubes.
4. Express the fact that there is a tetrahedron between two cubes.
5. Notice that it is not the case that every cube is in front of a dodecahedron. Say this.

Save your sentences as Sentences 6.16. Now let's change the world so that none of the above facts hold. We can do this by simply changing the medium cube into a dodecahedron. If your answers to 1–5 are correct, all of the sentences should now be false.

Problem$_\diamond$ 17. (Name that object) Open Carroll's World and Hercule's Sentences. The object is to try to figure out which objects have names, and what they are. You should be able to figure this out from the sentences, all of which are true. Once you have come to your conclusion, add the names to the objects and see if all the sentences are true. Save the modified world as World 6.17.

6.3 The step-by-step method of translation

When an English sentence contains more than one quantified noun phrase, translating it can become quite confusing unless it is approached in a very systematic manner. It often helps to have a number of intermediate steps, where quantified noun phrases are treated one at a time. For example, suppose we wanted to translate the sentence *Each cube is to the left of a tetrahedron.* Here, there are two quantified noun phrases: *each cube* and *a tetrahedron.* We can start by dealing with the first noun phrase, temporarily treating the complex phrase *is-to-the-left-of-a-tetrahedron* as a single unit. In other words, we can think of the sentence as a single quantifier sentence, on the order of *Each cube is small.* The translation would look like this:

$$\forall x\,(\mathsf{Cube}(x) \rightarrow x\ \textit{is-to-the-left-of-a-tetrahedron})$$

Of course, this is not a sentence in our language, so we need to translate the expression x *is-to-the-left-of-a-tetrahedron.* But we can think of this expression as a single quantifier sentence, at least if we pretend that x is a name. It has the same general form as the sentence ***b*** *is to the left of a tetrahedron*, and would be translated as:

$$\exists y\,(\mathsf{Tet}(y) \wedge \mathsf{LeftOf}(x, y))$$

Substituting this in the above, we get the desired translation of the original English sentence:

$$\forall x\,(\mathsf{Cube}(x) \rightarrow \exists y\,(\mathsf{Tet}(y) \wedge \mathsf{LeftOf}(x, y)))$$

This is exactly the sentence with which we began our discussion of mixed quantifiers.

This step-by-step process really comes into its own when there are lots of quantifiers in a sentence. It would be very difficult for a beginner to translate a sentence like *No cube to the right of a tetrahedron is to the left of a larger dodecahedron* in a single blow. Using the step-by-step method makes it straightforward. Eventually, though, you will be able to translate quite complex sentences, going through the intermediate steps in your head.

Exercises and Problems

Problem$^{\star}_{\diamond}$ **18.** (Using the step-by-step method of translation)

- Open Montague's Sentences. This file contains expressions that are halfway between English and first-order logic. Our goal is to edit this file until it contains translations of the following English sentences. You should read the English sentence below, make sure you understand how we got to the halfway point, and then complete the translation by replacing the hyphenated expression with a wff of first-order logic.

1. *Every cube is to the left of every tetrahedron.* [In the Sentence window, you see the halfway completed translation, together with some blanks that need to be replaced by wffs. Commented out below this, you will find an intermediate "sentence." Make sure you understand how we got to this intermediate stage of the translation. Then complete the translation by replacing the blank with

   $$\forall y\ (\mathsf{Tet}(y) \rightarrow \mathsf{LeftOf}(x, y))$$

   Once this is done, check to see if you have a well-formed sentence. Does it look like a proper translation of the original English? It should.]
2. *Every small cube is in back of a large cube.*
3. *Some cube is in front of every tetrahedron.*
4. *A large cube is in front of a small cube.*
5. *Nothing is larger than everything.*
6. *Every cube in front of every tetrahedron is large.*
7. *Everything to the right of a large cube is small.*
8. *Nothing in back of a cube and in front of a cube is large.*
9. *Anything with nothing in back of it is a cube.*
10. *Every dodecahedron is smaller than some tetrahedron.*

Save your sentences as Sentences 6.18.

- Open Peirce's World. Notice that all the English sentences are true in this world. Check to see that all of your translations are true as well. If they are not, see if you can figure out where you went wrong.
- Open Leibniz's World. Note that the English sentences 5, 6, 8, and 10 are true in this world, while the rest are false. Verify that your translations have the same truth values. If they don't, fix them.
- Open Ron's World. Here, the true sentences are 2, 3, 4, 5, and 8. Check that your translations have the right values, and correct them if they don't.

Problem$_\diamond$ 19. (More multiple quantifier sentences) Now, we will try translating some multiple quantifier sentences completely from scratch. You should try to use the step-by-step procedure.

- Start a new sentence file and translate the following English sentences.
  1. *Every tetrahedron is in front of every dodecahedron.*
  2. *No dodecahedron has anything in back of it.*
  3. *No tetrahedron is the same size as any cube.*

4. *Every dodecahedron is the same size as some cube.*
5. *Anything between two tetrahedra is a small cube.* [Note: This use of *two* really can be paraphrased using *between a tetrahedron and a tetrahedron.*]
6. *Every cube falls between two objects.*
7. *Every cube with something in back of it is small.*
8. *Every dodecahedron with nothing to its right is small.*
9. (⋆) *Every dodecahedron with nothing to its right has something to its left.*
10. *Any dodecahedron to the left of a cube is large.*

Save your sentences as Sentences 6.19.

- Open Bolzano's World. All of the above English sentences are true in this world. Verify that all your translations are true as well.
- Now open Ron's World. The English sentences 4, 5, 8, 9, and 10 are true, but the rest are false. Verify that the same holds of your translations.
- Open Claire's World. Here you will find that the English sentences 1, 3, 5, 7, 9, and 10 are true, the rest false. Again, check to see that your translations have the appropriate truth value.
- Finally, open Peano's World. Notice that only sentences 8 and 9 are true. Check to see that your translations have the same truth values.

6.4 Paraphrasing English

Some English sentences do not easily lend themselves to direct translation using the step-by-step procedure. With such sentences, however, it is often quite easy to come up with an English paraphrase that is amenable to the procedure. Consider, for example, *If a freshman takes a logic class, then he or she must be smart.* The step-by-step procedure does not work here. If we try to apply the procedure we would get something like:

$$\exists x(\mathsf{Freshman}(x) \land \exists y(\mathsf{LogicClass}(y) \land \mathsf{Takes}(x, y))) \rightarrow \mathsf{Smart}(x)$$

The problem is that this "translation" is not a sentence, since the last occurrence of x is free. However, we can paraphrase the sentences as *Every freshman who takes a logic class must be smart.* This is easily treated by the procedure, with the result being:

$$\forall x[(\mathsf{Freshman}(x) \land \exists y(\mathsf{LogicClass}(y) \land \mathsf{Takes}(x, y))) \rightarrow \mathsf{Smart}(x)]$$

> **Remember**
>
> In translating from English to FOL, the goal is to get a sentence that has the same meaning as the original. This sometimes requires changes in the surface form of the sentence.

Exercises and Problems

Problem$^{\star}_{\diamond}$ 20. (Sentences that need paraphrasing before translation)

- Translate the following sentences by first giving a suitable English paraphrase.
  1. *Only large objects have nothing in front of them.*
  2. *If a cube has something in front of it, then it's small.*
  3. *Every cube in back of a dodecahedron is also smaller than it.* [Warning: This is an example of what is known as a "donkey" sentence, following a notorious example *Every farmer who owns a donkey beats it.* What makes such a sentence a bit tricky is the existential noun phrase in the relative clause which serves as the antecedent of the pronoun "it" in the verb phrase. This combination in effect forces us to translate the existential noun phrase with a universal quantifier. First, the donkey sentence would be paraphrased as *For every farmer and every donkey, if the farmer owns the donkey, then he beats it.* This sentence clearly needs two universal quantifiers in its translation. Several of the sentences that follow are donkey sentences.]
  4. *If* **e** *is between two objects, then they are both small.*
  5. *If a tetrahedron is between two objects, then they are both small.*
  6. *Every dodecahedron is at least as large as every cube.* [Hint: Since we do not have anything corresponding to *as large as* in our language, you will first need to paraphrase this using *larger than* or *smaller than.*]
  7. *If a cube is to the right of a dodecahedron but not in back of it, then it is as large as the dodecahedron.*
  8. *No cube with nothing to its left is between two cubes.*
  9. *The only large cubes are* ***b*** *and* ***c***.
  10. *At most* ***b*** *and* ***c*** *are large cubes.* [Note: There is a significant difference between this sentence and the previous one. This one does not imply that b and c are large cubes, while the previous sentence does.]

  Save your sentences as Sentences 6.20.

- Open Ron's World. Recall that there are lots of hidden things in this world. Each of the above English sentences is true in this world, so the same should hold of your translations. Check to see that it does.
- Now open Bolzano's World. In this world, only sentences 3, 8, and 10 are true. Check that the same holds of your translations.
- Next open Wittgenstein's World. In this world, only the English sentences 5, 7, and 8 are true. Verify that your translations have the same truth values.

Problem$_\diamond$ 21. (Translation) Open Peirce's World. Look at it in 2-D to remind yourself of the hidden objects. Start a new sentence file where you will translate the following English sentences. Again, be sure to check each of your translations to see that it is indeed a true sentence.

1. *Everything is either a cube or a tetrahedron.*
2. *Every cube is to the left of every tetrahedron.*
3. *There are at least three tetrahedra.*
4. *Every small cube is in back of a large cube.*
5. *Every tetrahedron is small.*
6. *Every dodecahedron is smaller than some tetrahedron.* [Note: This is vacuously true in this world.]

Save the sentences as Sentences 6.21. Now let's change the world so that none of the English sentences are true. (We can do this by changing the large cube in front to a dodecahedron, the large cube in back to a tetrahedron, and deleting the two small tetrahedra in the far right column.) If your answers to 1–6 are correct, all of your translations should be false as well. If not, you have made a mistake in translation. Make further changes, and check to see that the truth values of your translations track those of the English sentences.

Problem$^{**}_\diamond$ 22. (More translations for practice) This exercise is just to give you more practice translating sentences of various sorts. They are all true in Skolem's World, in case you want to look while translating.

- Translate the following sentences.
  1. *Not all of the cubes are smaller than all of the tetrahedra.*
  2. *Not one of the cubes is to the right of anything.*
  3. *There is a dodecahedron unless there are at least two large objects.*
  4. *No cube with nothing in back of it is smaller than another cube.*
  5. *If any dodecahedra are small, then they are between two cubes.*

6. *If a cube is medium or is in back of something medium, then it has nothing to its right except for tetrahedra.*
7. *The further back a thing is, the larger it is.*
8. *Everything is the same size as something else.*
9. *Every cube has a tetrahedron of the same size to its right.*
10. *Nothing is the same size as two (or more) other things.*
11. *Nothing is between objects of shapes other than its own.*

Save your sentences as Sentences 6.22.

- Open Skolem's World. Notice that all of the above English sentences are true. Verify that the same holds of your translations.
- This time, rather than open other worlds, make changes to Skolem's World and see that the truth value of your translations track that of the English sentence. For example, consider sentence 5. Add a small dodecahedron between the front two cubes. The English sentence is still true. Is your translation? Now move the dodecahedron over between two tetrahedra. The English sentence is false. Is your translation? Now make the dodecahedron medium. The English sentence is again true. How about your translation?

Problem$^{\star}_{\diamond}$ 23. (More translations) The following English sentences are true in Gödel's World. Translate them, and make sure your translations are also true. Then modify the world in various ways, and check that your translations track the truth value of the English sentence.

1. *Nothing to the left of **a** is larger than everything to the left of **b**.*
2. *Nothing to the left of **a** is smaller than anything to the left of **b**.*
3. *The same things are left of **a** as are left of **b**.*
4. *Anything to the left of **a** is smaller than something that is in back of every cube to the right of **b**.*
5. *Every cube is smaller than some dodecahedron but no cube is smaller than every dodecahedron.*
6. *If **a** is larger than some cube then it is smaller than every tetrahedron.*
7. *Only dodecahedra are larger than everything else.*
8. *All objects with nothing in front of them are tetrahedra.*
9. *Nothing is between two objects which are the same shape.*
10. *Nothing but a cube is between two other objects.*
11. ***b** has something behind it which has at least two objects behind it.*
12. *More than one thing is smaller than something larger than **b***

Save your sentences as Sentences 6.23.

Problem 24. Using the symbols introduced in Table 1, page 23, translate the following into FOL. Do not introduce any additional names or predicates. Comment on any shortcomings in your translations.

1. *Every student gave a disk to some other student sometime or other.*
2. *Claire is not a student unless she has a disk.*
3. *No one ever owned both Folly and Silly at the same time.*
4. *No student erased every disk.*
5. *No one who owned a disk at 2:00 was angry.*
6. *No one gave Claire a disk this morning.* (Assume this was said on January 2, 1993.)
7. *If Max ever gave Claire a disk, she owned it then and he didn't.*
8. *You can't give someone something you don't own.*
9. *Max erased all of his disks before Claire erased any of her disks.*
10. *Max gave Claire a disk between 2:00 and 3:00. It was blank.*
11. *No student owned two disks at a time.*
12. *No student owned two disks until Claire did.*
13. *Anyone who owns a disk erases it sometime.*
14. *Anyone who owns a disk erases it sometime while they own it.*
15. *Only disks that aren't blank are erased.*

Problem 25. Using the symbols introduced in Table 1, page 23, translate the following into colloquial English. Assume that each of the sentences was asserted at 2 p.m. on January 2, 1993, and use this fact to make your translations more natural. For example, you could translate Owned(Max, Folly, 2:00) as *Max owns Folly.*

1. $\forall x[\mathsf{Student}(x) \rightarrow \exists z(\mathsf{Disk}(z) \wedge \mathsf{Owned}(x, z, 2{:}00))]$
2. $\exists x[\mathsf{Student}(x) \wedge \forall z(\mathsf{Disk}(z) \rightarrow \mathsf{Owned}(x, z, 2{:}00))]$
3. $\forall x \forall t[\mathsf{Gave}(\mathsf{Max}, x, \mathsf{Claire}, t) \rightarrow \exists y \exists t' \mathsf{Gave}(\mathsf{Claire}, x, y, t')]$
4. $\exists x[\mathsf{Owned}(\mathsf{Claire}, x, 2{:}00) \wedge \exists t(t < 2{:}00 \wedge \mathsf{Gave}(\mathsf{Max}, x, \mathsf{Claire}, t))]$
5. $\exists x \exists t(1{:}55 < t \wedge t < 2{:}00 \wedge \mathsf{Gave}(\mathsf{Max}, x, \mathsf{Claire}, t))$
6. $\forall y[\mathsf{Person}(y) \rightarrow \exists x \exists t(1{:}55 < t \wedge t < 2{:}00 \wedge \mathsf{Gave}(\mathsf{Max}, x, y, t))]$
7. $\exists z\{\mathsf{Student}(z) \wedge \forall y[\mathsf{Person}(y) \rightarrow \exists x \exists t(1{:}55 < t \wedge t < 2{:}00 \wedge \mathsf{Gave}(z, x, y, t))]\}$

Problem* 26. Translate the following into FOL. As usual, explaining the meanings of the names, predicates, and function symbols you use, and comment on any shortcomings in your translations.

1. *There's a sucker born every minute.*
2. *Whither thou goest, I will go.*
3. *Soothsayers make a better living in the world than truthsayers.*
4. *To whom nothing is given, nothing can be required.*

5. *If you always do right, you will gratify some people and astonish the rest.*
6. *There was a jolly miller once*
   Lived on the River Dee;
   He worked and sang from morn till night
   No lark more blithe than he.
7. *You should always except the present company.*
8. *Man is the only animal that blushes. Or needs to.*
9. *You can fool all of the people some of the time, and some of the people all of the time, but you can't fool all of the people all of the time.*
10. *Everybody loves a lover.*

6.5 Ambiguity and context sensitivity

There are a couple of things that make the task of translating between English and first-order logic difficult. One is the sparsity of primitive concepts in FOL. While this sparsity makes the language easy to learn, it also means that there are frequently no very natural ways of saying what you want to say. You have to try to find circumlocutions available with the resources at hand. While this is often possible in mathematical discourse, it is frequently impossible for ordinary English. (We will return to this matter later.)

The other thing that makes it difficult is that English is rife with ambiguities, whereas the expressions of first-order logic are unambiguous (at least if the predicates used are unambiguous). Thus, confronted with a sentence of English, we often have to choose one among many possible interpretations in deciding on an appropriate translation. Just which is appropriate usually depends on context.

The ambiguities become especially vexing with quantified noun phrases. Consider, for example, the following joke, taken from *Saturday Night Live*:

> "Every minute a man is mugged in New York City. We are going to interview him tonight."

What makes this joke possible is the ambiguity in the first sentence. The most natural reading would be translated by

$$\forall x(\mathsf{Minute}(x) \rightarrow \exists y(\mathsf{Man}(y) \land \mathsf{MuggedDuring}(y, x)))$$

But the second sentence forces us to go back and reinterpret the first in a rather unlikely way, one that would be translated by:

$$\exists y(\mathsf{Man}(y) \land \forall x(\mathsf{Minute}(x) \rightarrow \mathsf{MuggedDuring}(y, x)))$$

Notice that the reason the second translation is less likely is not determined by the form of the original sentence. You can find examples of the same form where the second reading is more natural. For example, suppose you have been out all day and, upon returning to your room, your roommate says, "Every ten minutes some guy from the registrar's office has called trying to reach you." Here it is the reading where the existential "some guy" is given wide scope that is most likely the one intended.

The problems of translation are much more difficult when we look at extended discourse, where more than one sentence comes in. To get a feeling for the difficulty, we start of with a couple of problems about extended discourse.

Exercises and Problems

Problem$^{\star}_{\diamond}$ 27. (Translating extended discourse)

- Open Reichenbach's World 1 and examine it. Check to see that all of the sentences in the following discourse are true in this world.

  There are (at least) two cubes. There is something between them. It is a medium dodecahedron. It is in front of a large dodecahedron. These two are left of a small dodecahedron. There are two tetrahedra.

  Translate this discourse into a single first-order sentence. Save your sentence as Sentences 6.27.1. Check to see that your translation is true. Now check to see that your translation is false in Reichenbach's World 2.

- Open Reichenbach's World 2. Check to see that all of the sentences in the following discourse are true in this world.

  There are two tetrahedra. There is something between them. It is a medium dodecahedron. It is in front of a large dodecahedron. There are two cubes. These two are left of a small dodecahedron.

  Translate this into a single first-order sentence, saving your answer as Sentences 6.27.2. Check to see that your translation is true. Now check to see that it is false in Reichenbach's World 1. However, note that the English sentences in the two discourses are in fact exactly the same; they have just been rearranged! The moral of this exercise is that the correct translation of a sentence into first-order logic (or any other language) can be very dependent on context.

Problem$^{\star}_{\diamond}$ 28. (Ambiguity) Create a sentence file Sentences 6.28 and translate the following sentences into first-order logic. Each of these sentences is ambiguous, so you should have at least two different translations of each.

1. *Every cube is between a pair of dodecahedra.*
2. *Every cube to the right of a dodecahedron is smaller than it is.*
3. *Cube* **a** *is not larger than every dodecahedron.*
4. *No cube is to the left of some dodecahedron.*
5. *(At least) two cubes are between (at least) two dodecahedra.*

Save the file. Now open Carroll's World. Which of your sentences are true in this world? You should find that exactly one translation of each sentence is true. Notice that if you had had the world in front of you when you did the translations, it would have been harder to see the ambiguity in the English sentences. The world would have provided a context that made one interpretation the natural one.

Problem 29. (Ambiguity and inference) Whether or not an argument is valid often hinges on how some ambiguous claim is taken. Here are some premises and a purported conclusion:

1. *Everyone admires someone who has red hair.*
2. *Anyone who admires himself is conceited.*
3. Conclusion: *Someone with red hair is conceited.*

Translate these into FOL twice, corresponding to the ambiguity in the first premise. Under one translation the conclusion follows. Prove it. Under the other, it does not. Describe a situation in which the premises are true (with this translation) but the conclusion is false.

Problem 30. (Ambiguity and inference) Here are some more premises and a purported conclusion:

1. *All that glitters is not gold.*
2. *This ring glitters.*
3. Conclusion: *This ring is not gold.*

Translate these into FOL twice, corresponding to the ambiguity in the first premise. Under one translation the conclusion follows. Prove it. Under the other, it does not. Describe a situation in which the premises are true (with this translation) but the conclusion is false.

Problem* 31. Give two translations of each of the following and discuss which is the most plausible reading, and why.

1. *Every senior in the class likes his or her computer, and so does the professor.* [Treat "the professor" as a name here and in the next sentence.]
2. *Every senior in the class likes his or her advisor, and so does the professor.*

3. *In some countries, every student must take an exam before going to college.*
4. *In some countries, every student learns a foreign language before going to college.*

6.6 Translations using function symbols

Intuitively, functions are a kind of relation. One's mother is one's mother because of a certain relationship you bear to one another. Similarly, $2 + 3 = 5$ because of a certain relationship between two, three, and five. Building on this intuition, it is not hard to see that anything that can be expressed in FOL with function symbols can also be expressed in a version of FOL where the function symbols have been replaced by relation symbols.

The basic idea can be illustrated easily. Let us use mother as a unary-function symbol, but MotherOf as a *binary* relation symbol. Thus, for example, mother(Max) = Nancy and MotherOf(Nancy, Max) both state that Nancy is the mother of Max.

The basic claim is that anything we can say with the function symbol we can say in some other way using the relation symbol. As an example, here is a simple sentence using the function symbol:

$$\forall x\ \mathsf{OlderThan}(\mathsf{mother}(x), x)$$

It expresses the claim that a person's mother is always older than the person. To express the same thing with the relation symbol, we might write

$$\forall x \exists y [\mathsf{MotherOf}(y, x) \land \mathsf{OlderThan}(y, x)]$$

Actually, one might wonder whether the second sentence quite manages to express the claim made by the first, since all it says is that everyone has at least one mother that is older than they are. One might prefer something like

$$\forall x \forall y [\mathsf{MotherOf}(y, x) \rightarrow \mathsf{OlderThan}(y, x)]$$

This says that every mother of everyone is older than they are. But this too seems somewhat deficient. A still better translation would be to conjoin one of the above sentences with the following two sentences which, together, assert that the relation of being the mother of someone is functional. Everyone has at least one, and everyone has at most one.

$$\forall x \exists y\ \mathsf{MotherOf}(y, x)$$

and

$$\forall x \forall y \forall z [(\mathsf{MotherOf}(y, x) \land \mathsf{MotherOf}(z, x)) \rightarrow y = z]$$

We will study this sort of thing much more in the next chapter, where we will see that these two sentences can jointly be expressed by one rather opaque sentence:

$$\forall x \exists y[\mathsf{MotherOf}(y, x) \land \forall z[\mathsf{MotherOf}(z, x) \to y = z]]$$

And, if we wanted to, we could then incorporate our earlier sentence and express the first claim by means of the horrendous looking:

$$\forall x \exists y[\mathsf{MotherOf}(y, x) \land \mathsf{OlderThan}(y, x) \land \forall z[\mathsf{MotherOf}(z, x) \to y = z]]$$

By now it should be clearer why function symbols are so useful. Look at all the connectives and additional quantifiers that have come into translating our very simple sentence

$$\forall x\ \mathsf{OlderThan}(\mathsf{mother}(x), x)$$

Remember

Anything that you can express using an n-ary function symbol can also be expressed using an $n+1$-ary relation symbol, plus the identity predicate, but at a considerable cost in terms of the complexity of the sentences used.

In this section we present some problems to give you practice translating sentences from English into FOL, problems that will show you why it is nice to have function symbols around.

Exercises and Problems

Problem 32. Translate the following sentences into FOL twice, once using the function symbol mother, once using the relation symbol MotherOf.

1. *Claire's mother is older than Max's mother.*
2. *Everyone's mother's mother is older than Melanie.*
3. *Someone's mother's mother is younger than Mary.*

Problem 33. Translate the following into a version of FOL that has function symbols height and father, the predicate $>$, and names for Max and Claire.

1. Max's father is taller than Max but not taller than Claire's father.
2. Someone is taller than Claire's father.
3. Everyone is taller than someone else.
4. No one is taller than himself.
5. Everyone who is taller than Claire is taller than Max.
6. Everyone who is shorter than Claire is shorter than someone who is shorter than Max's father.

Problem 34. Using the first-order language of arithmetic described earlier, express each of the following in FOL.

1. Every number is either 0 or greater than 0.
2. The sum of any two numbers greater than 1 is smaller than the product of the same two numbers.
3. Every number is even. [This is false, of course.]
4. If $x^2 = 1$ then $x = 1$.
5. ⋆ For any number x, if $ax^2 + bx + c = 0$ then either $x = \frac{-b+\sqrt{b^2-4ac}}{2a}$ or $x = \frac{-b-\sqrt{b^2-4ac}}{2a}$. In this problem use a, b, c as constant symbols, but treat x as a variable, as usual in algebra.

6.7 Methods of proof involving mixed quantifiers

There are really no new methods of proof that apply specifically to sentences with mixed quantifiers. But the introduction of mixed quantifiers forces us to be more explicit about some subtleties in the informal methods we have already introduced. The subtleties have to do with the methods that introduce new names into a proof: existential instantiation, general conditional proof, and universal generalization. It turns out that problems can arise from the interaction of these methods of proof.

Let us begin by illustrating the problem. Consider the following two sentences.

1. ∀x[Boy(x) → ∃y(Girl(y) ∧ Likes(x, y))]
2. ∃y[Girl(y) ∧ ∀x(Boy(x) → Likes(x, y))]

Assume that the domain of discourse is the set of children in Claire's and Max's kindergarten. Thus the first sentence says every boy in the class likes some girl or other, while the second says that there is some girl who is liked by every boy. You will realize by now that the second logically implies the first, but not vice versa. Let's start by giving a proof that the second does indeed imply the first.

> **Proof:** To prove that (2) logically implies (1), we begin by assuming (2). Thus, at least one girl is liked by every boy. Let us introduce the new name "c" for one of these popular girls. To prove (1) we will use general conditional proof. Assume that d is any boy in the class. We want to prove that d likes some girl. But every boy likes c, so d likes c. Thus d likes some girl, by existential generalization. Since d was an arbitrarily chosen boy, (1) is true. So (1) follows from (2).

This is a perfectly legitimate proof. The problem we want to illustrate, however, is the superficial similarity between the above proof and the following incorrect "proof" that (1) implies (2). Obviously, since (1) does not imply (2), we can't really prove that it does. But the following appears to do just that.

Pseudo-proof: Assume (1), that is, that every boy likes some girl or other. Let e be any boy in the domain. By (1), e likes some girl. Let us introduce the new name "f" for some girl that e likes. Since the boy e was chosen arbitrarily, we conclude that every boy likes f, by general conditional proof. But then, by existential generalization, we have the desired result, namely, that some girl is liked by every boy.

This reasoning is fallacious. Seeing why it is fallacious is extremely important, if we are to avoid missteps in reasoning. The problem centers on our conclusion that every boy likes f. Recall how the name "f" came into the proof. We knew that e, being one of the boys, liked some girl, and we chose one of those girls and dubbed her "f." This choice of a girl depends crucially on which boy e we are talking about. If e was Max or Alex, we could have picked Claire and dubbed her "f." But if e was Eric, we couldn't pick Claire. Eric likes one of the girls, but not Claire.

The problem is this. Recall that in order to conclude a universal claim based on reasoning about a single individual, it is imperative that we not appeal to anything specific about that individual. But after we give the name "f" to one of the girls that e likes, any conclusion we come to about e and f may well violate this imperative. We can't be positive that it would apply equally to all the boys.

Stepping back from this particular example, the upshot is that we cannot conclude a general conditional proof of $\forall x[P(x) \rightarrow Q(x)]$ from $P(c)$ if $Q(c)$ mentions a specific individual whose choice depended on the individual c. In practice, the best way to insure that no such individual is specifically mentioned is to insist that $Q(c)$ not contain *any* name that was introduced by existential instantiation after the assumption of $P(c)$.

A similar restriction must be placed on the use of universal generalization. Recall that universal generalization involves the introduction of a new constant, say c, standing for an arbitrary member of the domain of discourse. We said that if we could prove a sentence $S(c)$, we could then conclude $\forall x S(x)$. However, we must now add the restriction that $S(c)$ not contain any constant introduced by existential instantiation after the introduction of the constant c. This restriction prevents invalid proofs like the following.

Pseudo-proof: Assume $\forall x \exists y R(x, y)$. We will show that, ignoring the above restriction, we can "prove" $\exists y \forall x R(x, y)$. We begin by taking c as a name for an arbitrary member of the domain. By universal instantiation, we get $\exists y R(c, y)$. Let d be such that $R(c, d)$. Since c stands for an arbitrary object,

> we have $\forall x R(x, d)$. Hence, by existential generalization, we get $\exists y \forall x R(x, y)$.

Can you spot the fallacious step in this proof? The problem is that we generalized from $R(c, d)$ to $\forall x R(x, d)$. But the constant d was introduced by existential instantiation (though we did not say so explicitly) after the constant c was introduced. Hence, the choice of the object d depends on which object c we are talking about. The subsequent universal generalization is just what our restriction rules out.

Let us now give a refined statement of our methods of proof.

Remember

Let $S(x)$, $P(x)$, and $Q(x)$ be wffs.

1. Existential Instantiation: If you have proven $\exists x S(x)$ then you may choose a *new* constant symbol c and assume $S(c)$.
2. General Conditional: If you want to prove $\forall x[P(x) \rightarrow Q(x)]$ then you may choose a *new* constant symbol c, assume $P(c)$, and prove $Q(c)$, making sure that Q does not contain any names introduced by existential instantiation after the assumption of $P(c)$.
3. Universal Generalization: If you want to prove $\forall x S(x)$ then you may choose a new constant symbol c and prove $S(c)$, making sure that $S(c)$ does not contain any names introduced by existential instantiation after the introduction of c.

We illustrate the correct uses of these rules with a famous example, one that goes back at least to Euclid. We will prove that there are infinitely many primes.

We cannot directly express this claim in FOL, but we can express something equivalent to it. Namely, we can express that there is no largest prime. In FOL, we might put it this way:

$$\forall x \exists y [y \geq x \land \text{Prime}(y)]$$

Here the intended domain of discourse is the natural numbers, of course.

> **Proof:** We see that this sentence is a mixed quantifier sentence, of just the sort we have been examining. To prove it, we let n be an arbitrary natural number and try to prove that there exists a prime number at least as large as n. To prove this, let k be the product of all the prime numbers less than n. Thus each prime less than n divides k without remainder. So now let $m = k+1$. Each prime less than n divides m with remainder 1. But we know that m can be factored into primes. Let p be one of these primes. Clearly, by the earlier observation, p must be

greater than or equal to n. Hence, by existential generalization, we see that there does indeed exist a prime number greater than or equal to n. But n was arbitrary, so we have established our result.

Notice the order of the last two steps. Had we violated the new condition on the application of general conditional proof to conclude that p is a prime number greater than or equal to every natural number, we would have obtained a patently false result.

Here, by the way, is a closely related conjecture, called the Twin Prime Conjecture. No one knows whether it is true or not.

$$\forall x \exists y [y > x \land \mathsf{Prime}(y) \land \mathsf{Prime}(y + 2)]$$

We now turn to some problems and exercises involving proofs with mixed quantifiers.

Exercises and Problems

Problem 35. What, if anything, is wrong with the following proof that there is a number greater than every other number? Explain in terms of the above discussion.

> Every number is less than some other number, for example, x is less than $x + 1$. So let n be an arbitrary number. Then n is less than some other number. Let m be such a number. Thus $n \leq m$. But n is an arbitrary number, so every number is less or equal m. Hence there is a number that is greater than every other number.

Problem 36. Below is a purported proof that:

$$\forall x[\mathsf{Person}(x) \rightarrow \forall z(\mathsf{Person}(z) \rightarrow \exists y\ \mathsf{GivesTo}(x, y, z))]$$

is a logical consequence of:

$$\forall x[\mathsf{Person}(x) \rightarrow \exists y \forall z[\mathsf{Person}(z) \rightarrow \mathsf{GivesTo}(x, y, z)]]$$

Where does it go wrong, if it does? Explain the validity or invalidity of the use of general conditional proof in terms of the above discussion.

> Let us assume the premise and prove the conclusion. Let b be an arbitrary person in the domain of discourse. We need to prove
>
> $$\forall z(\mathsf{Person}(z) \rightarrow \exists y\ \mathsf{GivesTo}(b, y, z))$$
>
> Let c be an arbitrary person in the domain of discourse. We need to prove
>
> $$\exists y\ \mathsf{GivesTo}(b, y, c)$$
>
> But this follows directly from our premise, since there is something that b gives to everyone.

Problem 37. Here is a purported proof that there is at most one object in the domain of discourse. Explain where it goes wrong, if it does.

> Suppose, by way of proof by contradiction, that there is more than one object in the domain of discourse. Let c be any one of these objects. Then there is some other object d, so that $d \neq c$. But since c was arbitrary, $\forall x(d \neq x)$. But then, by universal instantiation, $d \neq d$. But $d = d$, so we have our contradiction. Hence there can be at most one object in the domain of discourse.

Problem$_\diamond$ 38. Consider the following premises:

1. $\forall x[\text{Cube}(x) \rightarrow \exists y\, \text{LeftOf}(x, y)]$
2. $\neg\exists x\exists z[\text{Cube}(x) \land \text{Cube}(z) \land \text{LeftOf}(x, z)]$
3. $\exists x\exists y[\text{Cube}(x) \land \text{Cube}(y) \land x \neq y]$

Is the following a logical consequence of these premises?

$$\exists x\exists y\exists z[\text{BackOf}(y, z) \land \text{LeftOf}(x, z)]$$

If so, give a proof. If not, build a world where the premises are true but the conclusion is false. Save it as World 6.38.

Problem$_\diamond$ 39. Use the premises of Problem 38. Does it follow that $\exists x\, \neg\text{Cube}(x)$? If so, prove it. If not, build a world where the premises are all true but the conclusion is false. Save it as World 6.39.

Problem$_\diamond$ 40. Use the premises of Problem 38. Does it follow that $\exists x\exists y(x \neq y \land \neg\text{Cube}(x) \land \neg\text{Cube}(y))$? If so, prove it. If not, build a world where the premises are all true but the conclusion is false. Save it as World 6.40.

Problem$_\diamond$ 41. Is the following logically true?

$$\exists x[\text{Cube}(x) \rightarrow \forall y\, \text{Cube}(y)]$$

If so, prove it. If not, build a world where it is false. Save it as World 6.41.

Problem 42. Consider the following argument.

1. *Whenever a student owns no disks, they are angry.*
2. *Claire, a student, was not angry at 2:00, but she was angry at 2:05.*
3. Conclusion: *Claire gave away at least one disk between 2:00 and 2:05.*

Does the conclusion follow from the premises? If so, give an informal proof. If not, describe possible circumstances that make the premises true but the conclusion false. You may assume that if someone owns a disk at some time and then later does not own a disk, then the person must have given the disk away sometime between the two times.

Problem 43. Consider the following argument.

1. *No student owned two disks until after 3:00.*
2. *Claire owned Folly at 2:00.*
3. *Claire owned Silly at 2:30.*
4. Conclusion: *If Claire did not give anyone a disk before 3:00, then she is not a student.*

Does the conclusion follow from the premises? If so, give an informal proof. If not, describe circumstances that would make the premises true but the conclusion false. You may use the same assumption as in the previous problem.

Problem* 44. Call a natural number a *near prime* if its prime factorization contains at most two distinct primes. The first number which is not a near prime is $2 \times 3 \times 5 = 30$. Prove

$$\forall x \exists y [y > x \wedge \neg \mathsf{NearPrime}(y)]$$

You may appeal to our earlier result that there is no largest prime.

6.8 Formal proofs

A formal proof system like $\mathcal{F}$ is intended as a mathematical model of rigorous reasoning. Because of this, it is designed to make perspicuous the key inferential moves and keep track of the various dependencies between steps.

This is nicely illustrated with the quantifier rules. It turns out that the new restrictions that we added to our informal methods are already ensured by the mechanisms built into $\mathcal{F}$. The formal rules we gave in the previous chapter do not need to be changed or amplified in any way. They are already set up to avoid the sorts of problems we ran into.

To see how $\mathcal{F}$ prevents the misapplication of universal generalization, let's formalize the fallacious proof given on page 173 and see how it violates the rules.

1. $\forall x \exists y R(x, y)$	
[c]	
2. $\exists y R(c, y)$	$\forall$ Elim: 1
[d]	
3. $R(c, d)$	
4. $R(c, d)$	Reit: 3
5. $R(c, d)$	$\exists$ Elim: 2, [d]-4
6. $\forall x R(x, d)$	$\forall$ Intro: [c]-5
7. $\exists y \forall x R(x, y)$	$\exists$ Intro: 6

The incorrect step here is step 5, since it violates the restriction on existential elimination that requires the constant d to appear only in the subproof where it is introduced. If we *could* make that move, then the rest of the proof would be fine.

Let's contrast the above, faulty proof, with a genuine proof that $\forall x \exists y R(x, y)$ follows from $\exists y \forall x R(x, y)$.

1. $\exists y \forall x R(x, y)$	
[d]	
2. $\forall x R(x, d)$	
[c]	
3. $R(c, d)$	$\forall$ Elim: 2
4. $\exists y R(c, y)$	$\exists$ Intro: 3
5. $\forall x \exists y R(x, y)$	$\forall$ Intro: [c]-4
6. $\forall x \exists y R(x, y)$	$\exists$ Elim: 1, [d]-5

Notice that in this proof, unlike the earlier one, both constant symbols c and d are properly sequestered within the subproofs where they are introduced. Therefore the quantifier rules have been applied properly.

$\mathcal{F}'$: Further extensions to $\mathcal{F}$

In section 4.6, we extended the Fitch system with several rules that were convenient but not strictly speaking necessary. Once we add quantifier rules to the system, there are several further conveniences that can be added to $\mathcal{F}'$.

The first addition is a formal counterpart of the method of general conditional proof. The schematic statement of this rule runs as follows:

General Conditional Proof (Gen Cond):

[c] P(c)	Where c does not occur outside the subproof where it is introduced.
⋮	
Q(c)	
▷ $\forall x (P(x) \rightarrow Q(x))$	

Within the context of giving formal proofs, this rule does not save us that much effort. After all, any proof that uses this rule can easily be transformed into a very similar proof using universal introduction and conditional introduction:

```
| | [c]
| |--
| | | P(c)
| | |--
| | |  :
| | | Q(c)
| | P(c) → Q(c)
| ∀x(P(x) → Q(x))
```

Nonetheless, the rule of general conditional proof allows our formal proofs to mirror informal reasoning more accurately.

The second modification to $\mathcal{F}'$ will allow us to give neater proofs involving statements with strings of universal quantifiers or strings of existential quantifiers. Suppose, for example, we wanted to prove a claim of the form $\forall x \forall y \forall z\, S(x, y, z)$. In informal reasoning, we would begin by saying "Let a, b, and c be arbitrary objects in the domain." In the system $\mathcal{F}$, however, a proof of this sort would require a separate nested subproof for each of the three quantifiers.

We can easily remedy this inconvenience by slight modifications to the quantifier rules, including that of general conditional proof. The modified version of universal introduction can be given schematically as follows:

Universal Introduction (∀ Intro):

```
  | | [c1, ..., cn]
  | |--
  | |  :
  | | P(c1, ..., cn)
▷ | ∀x1 ... ∀xn P(x1, ..., xn)
```

Where the c_i's are distinct and do not occur outside the subproof where they are introduced.

In the statement of this rule, $c_1 \ldots c_n$ can be any string of distinct constant symbols; $x_1 \ldots x_n$ can be any string of distinct variables; and $P(c_1, \ldots, c_n)$ is the result of replacing each free occurrence of x_i with c_i. In applying this rule, we do not want to have to worry about the exact order in which the constants appear at the top of the subproof, and so we will say that an application is correct if some reordering of the boxed constants fits the schema.

Here is the corresponding modified elimination rule:

Universal Elimination (∀ Elim):

$$\begin{array}{l|l} & \forall x_1 \ldots \forall x_n S(x_1, \ldots, x_n) \\ & \vdots \\ \triangleright & S(c_1, \ldots, c_n) \end{array}$$

In this rule, unlike in universal generalization, the c_i's need not be distinct. For example, we can substitute the same constant for two of the variables.

Here is a simple example of how these two rules work together. We will prove that $\forall z \forall y \forall x\, R(z, x, y, z)$ follows from $\forall w \forall x \forall y \forall z\, R(w, x, y, z)$.

$$\begin{array}{l l} 1.\ \forall w \forall x \forall y \forall z\, R(w, x, y, z) & \\ \quad \boxed{a, b, c} & \\ \quad 2.\ R(c, a, b, c) & \forall\ \text{Elim: } 1 \\ 3.\ \forall z \forall y \forall x\, R(z, x, y, z) & \forall\ \text{Intro: } \boxed{a, b, c}\text{-2} \end{array}$$

The remaining quantifier rules get modified in analogous ways. We list here the schematic forms.

Existential Introduction (∃ Intro):

$$\begin{array}{l|l} & S(c_1, \ldots, c_n) \\ & \vdots \\ \triangleright & \exists x_1 \ldots \exists x_n S(x_1, \ldots, x_n) \end{array}$$

Here again, as in universal elimination, the c_i's need not be distinct constants.

Existential Elimination (∃ Elim):

$$\begin{array}{l|l} & \exists x_1 \ldots \exists x_n S(x_1, \ldots, x_n) \\ & \vdots \\ & \quad \boxed{c_1, \ldots, c_n} \\ & \quad S(c_1, \ldots, c_n) \\ & \quad \vdots \\ & \quad Q \\ \triangleright & Q \end{array}$$

Where the c_i's are distinct and do not occur outside the subproof where they are introduced.

General Conditional Proof (Gen Cond):

$$
\begin{array}{l|l}
 & \boxed{c_1, \ldots, c_n} \\
 & P(c_1, \ldots, c_n) \\
 & \vdots \\
 & Q(c_1, \ldots, c_n) \\
\triangleright & \forall x_1 \ldots \forall x_n (P(x_1, \ldots, x_n) \to Q(x_1, \ldots, x_n))
\end{array}
$$

Where the c_i's are distinct and do not occur outside the subproof where they are introduced.

Unless otherwise noted, you should feel free to use the rules of $\mathcal{F}'$ in the problems that follow.

Exercises and Problems

Problem$_\diamond$ 45. Does $\exists x \exists y \neg \mathsf{LeftOf}(x, y)$ follow from $\exists x \neg \mathsf{LeftOf}(x, x)$? If so, give a formal proof. If not, build a world where the premise is true and the conclusion is false. Save it as World 6.45.

Problem 46. Give a formal proof of the following:

1. $\forall x \exists y\, \mathsf{Likes}(x, y)$ from the premise $\forall x \forall y\, \mathsf{Likes}(x, y)$.
2. $\exists x\, \mathsf{Cube}(x)$ from the premises $\forall x(\mathsf{Small}(x) \to \mathsf{Cube}(x))$, $\exists x \neg \mathsf{Cube}(x) \to \exists x\, \mathsf{Small}(x)$.
3. $\exists x\, \mathsf{Likes}(x, \mathsf{Carl})$ from $\mathsf{Likes}(\mathsf{Carl}, \mathsf{Max})$ and $\forall x[\exists y(\mathsf{Likes}(y, x) \lor \mathsf{Likes}(x, y)) \to \mathsf{Likes}(x, x)]$.
4. $\forall x \exists y\, \mathsf{Likes}(x, y)$ from the premises $\forall x \forall y[\mathsf{Likes}(x, y) \to \mathsf{Likes}(y, x)]$ and $\exists x \forall y\, \mathsf{Likes}(x, y)$.

Problem 47. Give a formal proof in the unmodified system $\mathcal{F}$ of the example worked on page 180.

Problem* 48. Give a formal proof of $\exists x(\mathsf{P}(x) \to \forall y \mathsf{P}(y))$ without premises. [Hint: This is a case where it will be helpful to give an informal proof before you try to construct the formal proof.]

6.9 Prenex form

When we translate complex sentences of English into FOL, it is common to end up with sentences where the quantifiers and connectives are all scrambled together. This is usually due to the way that the translations of complex noun phrases of English use both quantifiers and connectives:

$$\forall x(\mathsf{P}(x) \to \ldots)$$
$$\exists x(\mathsf{P}(x) \land \ldots)$$

As a result, the translation of (the most likely reading of) a sentence like *Every cube to the left of a tetrahedron is in back of a dodecahedron* ends up looking like:

$$\forall x[(\mathsf{Cube}(x) \land \exists y(\mathsf{Tet}(y) \land \mathsf{LeftOf}(x, y))) \rightarrow \exists y(\mathsf{Dodec}(y) \land \mathsf{BackOf}(x, y))]$$

While this is the most natural translation of our sentence, there are situations where it is not the most convenient one. It is sometimes important that we be able to rearrange sentences like this so that all the quantifiers are out in front, all the connectives in back. Such a sentence is said to be in *prenex form*, since all the quantifiers come first.

Stated more precisely, a wff is in *prenex normal form* if either it contains no quantifiers at all, or else is of the form

$$Q_1 v_1 Q_2 v_2 \ldots Q_n v_n P$$

where each Q_i is either $\forall$ or $\exists$, each v_i is some variable, and the wff P is quantifier-free.

There are several reasons one might want to put sentences into prenex form. One is that it gives you a nice measure of the logical complexity of the sentences. What turns out to matter is not so much the number of quantifiers, as the number of times you get a flip from $\forall$ to $\exists$ or the other way round. The more of these so-called "alternations," the more complex the sentence is, logically speaking. Another reason is that this prenex form is quite analogous to the conjunctive normal form for quantifier-free wffs we studied earlier. And like that normal form, it is used extensively in automated theorem proving.

It turns out that every sentence is logically equivalent to one (in fact many) in prenex form. In this section we will present some rules for carrying out this transformation. When we apply the rules to our earlier example, we will get:

$$\forall x \forall y \exists z[(\mathsf{Cube}(x) \land \mathsf{Tet}(y) \land \mathsf{LeftOf}(x, y)) \rightarrow (\mathsf{Dodec}(z) \land \mathsf{BackOf}(x, z))]$$

To arrive at this sentence, we did not just blindly pull quantifiers out in front. If we had, it would have come out all wrong. There are two problems. One is that the first $\exists y$ in the original sentence is, logically speaking, inside a $\neg$. (To see why, replace $\rightarrow$ by its definition in terms of $\neg$ and $\lor$.) Applying the DeMorgan laws for quantifiers tells us that it will end up being a universal quantifier. Another problem is that the original sentence has two quantifiers that bind the variable y. There is no problem with this, but if we pull the quantifiers out front, there is suddenly a clash. So we must first change one of the ys to some other variable, say z.

We have already seen some of the logical equivalences that are needed for putting sentences in prenex form. For example, the DeMorgan rules tell us how to pull a quantifier out in front of $\neg$. The rule about changing variables is important, as we have just seen. The other rules are summarized below. You should make sure that you understand why each of them holds.

Remember

Any sentence is logically equivalent to one in prenex form. One can obtain the prenex form using quantifier rules given earlier, together with the following equivalences.

$$
\begin{array}{rcll}
\forall x P \wedge \forall x Q & \Leftrightarrow & \forall x[P \wedge Q] & \\
\exists x P \vee \exists x Q & \Leftrightarrow & \exists x[P \vee Q] & \\
\forall x P \vee Q & \Leftrightarrow & \forall x[P \vee Q] & \text{if } x \text{ not free in } Q \\
\exists x P \wedge Q & \Leftrightarrow & \exists x[P \wedge Q] & \text{if } x \text{ not free in } Q \\
Q & \Leftrightarrow & \forall x Q & \text{if } x \text{ not free in } Q \\
Q & \Leftrightarrow & \exists x Q & \text{if } x \text{ not free in } Q
\end{array}
$$

In order to apply these rules to sentences with $\rightarrow$ or $\leftrightarrow$, one needs to either replace these symbols with equivalent versions using $\neg, \vee$ and $\wedge$, or else derive some similar rules for these symbols. Here is a step-by-step transformation of our original sentence into the one in prenex form given above. We have abbreviated the predicates in order to make it easier to read.

$$
\begin{array}{ll}
\forall x[(C(x) \wedge \exists y(T(y) \wedge L(x,y))) \rightarrow \exists y(D(y) \wedge B(x,y))] & \Leftrightarrow \\
\forall x[\neg(C(x) \wedge \exists y(T(y) \wedge L(x,y))) \vee \exists y(D(y) \wedge B(x,y))] & \Leftrightarrow \\
\forall x[\neg\exists y(C(x) \wedge T(y) \wedge L(x,y)) \vee \exists y(D(y) \wedge B(x,y))] & \Leftrightarrow \\
\forall x[\forall y\neg(C(x) \wedge T(y) \wedge L(x,y)) \vee \exists y(D(y) \wedge B(x,y))] & \Leftrightarrow \\
\forall x[\forall y\neg(C(x) \wedge T(y) \wedge L(x,y)) \vee \exists z(D(z) \wedge B(x,z))] & \Leftrightarrow \\
\forall x\forall y[\neg(C(x) \wedge T(y) \wedge L(x,y)) \vee \exists z(D(z) \wedge B(x,z))] & \Leftrightarrow \\
\forall x\forall y[\exists z\neg(C(x) \wedge T(y) \wedge L(x,y)) \vee \exists z(D(z) \wedge B(x,z))] & \Leftrightarrow \\
\forall x\forall y\exists z[\neg(C(x) \wedge T(y) \wedge L(x,y)) \vee (D(z) \wedge B(x,z))] & \Leftrightarrow \\
\forall x\forall y\exists z[(C(x) \wedge T(y) \wedge L(x,y)) \rightarrow (D(z) \wedge B(x,z))] &
\end{array}
$$

As important as these rules is the observation that some superficially similar transformations are not valid. In particular:

$$
\begin{array}{rcl}
\forall x P \vee \forall x Q & \not\Leftrightarrow & \forall x[P \vee Q] \\
\exists x P \wedge \exists x Q & \not\Leftrightarrow & \exists x[P \wedge Q]
\end{array}
$$

Exercises and Problems

Problem 49. Derive the following rules from those given earlier, by replacing $\rightarrow$ by its definition in terms of $\vee$ and $\neg$.

$$
\begin{array}{rcll}
\forall x P \rightarrow Q & \Leftrightarrow & \exists x[P \rightarrow Q] & \text{if } x \text{ not free in } Q \\
\exists x P \rightarrow Q & \Leftrightarrow & \forall x[P \rightarrow Q] & \text{if } x \text{ not free in } Q \\
P \rightarrow \forall x Q & \Leftrightarrow & \forall x[P \rightarrow Q] & \text{if } x \text{ not free in } P \\
P \rightarrow \exists x Q & \Leftrightarrow & \exists x[P \rightarrow Q] & \text{if } x \text{ not free in } P
\end{array}
$$

Problem$_\diamond$ 50. (Some invalid quantifier manipulations) We remarked above on the invalidity of some quantifier manipulations that are superficially similar to the valid ones. In fact, in both cases one side is a logical consequence of the other side, but not vice versa. We will illustrate this. Build a world in which (1) and (3) below are true, but (2) and (4) are false.

1. ∀x[Cube(x) ∨ Tet(x)]
2. ∀xCube(x) ∨ ∀xTet(x)
3. ∃xCube(x) ∧ ∃xSmall(x)
4. ∃x[Cube(x) ∧ Small(x)]

Save your world as as World 6.50.

Problem$_\diamond$ 51. (Putting sentences in Prenex form) Open Jon Russell's Sentences. You will find ten sentences, at the odd numbered positions. Write a prenex form of each sentence in the space below it. Save your sentences as Sentences 6.51. Open a few worlds, and make sure that your prenex form has the same truth value as the sentence above it.

Problem* 52. Translate the following two sentences into FOL:

1. *If everyone comes to the party, I will have to buy more food.*
2. *There is someone such that if he comes to the party, I will have to buy more food.*

The natural translations of these turn out to have forms that are equivalent, according to the first equivalence in Problem 49. But clearly the English sentences do not mean the same thing. Explain what is going on here. Are the natural translations really correct?

7

Some Specific Uses of Quantifiers

In this chapter we look at some specific uses of quantifiers. We begin with the use that justifies the name, where quantifiers are used to express numerical claims.

7.1 Making numerical claims

By a "numerical claim" we mean one that says that there are a certain number of objects with some property or other. In earlier problems, we have already come across some simple numerical claims. The problems in this section will help you recognize numerical claims when you come across them in our first-order language, as well as learn how to express them yourself.

Perhaps the single most important thing to remember is that distinct names do not necessarily refer to distinct objects, and distinct variables need not vary over distinct objects. For example, both of the following sentences can be made true in a world with one object.

$$\mathsf{Cube(a)} \land \mathsf{Small(a)} \land \mathsf{Cube(b)}$$

$$\exists \mathsf{x} \exists \mathsf{y}[\mathsf{Cube(x)} \land \mathsf{Small(x)} \land \mathsf{Cube(y)}]$$

In order to say that there are really two *distinct* objects, you must find a way to guarantee that they are different. For example, either of the following would do:

$$\mathsf{Cube(a)} \land \mathsf{Small(a)} \land \mathsf{Cube(b)} \land \mathsf{Large(b)}$$

$$\exists \mathsf{x} \exists \mathsf{y}[\mathsf{Cube(x)} \land \mathsf{Small(x)} \land \mathsf{Cube(y)} \land \mathsf{LeftOf(x, y)}]$$

The simplest way, though, is just to say that they are distinct:

$$\exists \mathsf{x} \exists \mathsf{y}[\mathsf{Cube(x)} \land \mathsf{Small(x)} \land \mathsf{Cube(y)} \land \mathsf{x} \neq \mathsf{y}]$$

This sentence asserts that there are at least two distinct cubes, at least one of which is small.

The uniqueness quantifier $\exists!$

One kind of numerical claim is so common that a special notational convention has arisen around it. It is where we assert that there is exactly one object satisfying some condition $\mathsf{P(x)}$. This can be expressed in FOL as follows:

$$\exists x[P(x) \land \forall y(P(y) \rightarrow y = x)]$$

as long as y does not occur already in the wff $\mathsf{P(x)}$. This is abbreviated as $\exists!x P(x)$. It is read "there is a unique x such that $P(x)$." Note that this is not a new quantifier, since wffs in which it occurs are just abbreviations for longer wffs involving the old quantifiers.

In a similar vein, we will sometimes write $\exists!^{n} x P(x)$ for the FOL sentence asserting that there are exactly n objects satisfying $\mathsf{P(x)}$. This is read "There are exactly n objects x satisfying $P(x)$." In both cases, it is important to remember that this notation is not part of the official language of FOL, but an abbreviation for a much longer FOL expression. Tarski's World does not recognize these abbreviations, so in using the program you will have to write them out in full.

Exercises and Problems

Exercise$_\diamond$ 1. (Numerical sentences) Open Whitehead's Sentences.

1. The first sentence says that there are at least two objects, and the second sentence says that there are at most two objects. (Do you see how they manage to say these things?) Build a model where the first two sentences are both true.
2. Sentence 3 is the conjunction of the first two. Hence it asserts, in one sentence, that there are exactly two objects. Check to see that it is true in the world you have just built.
3. The fourth sentence is in fact equivalent to the third sentence. It is a shorter way of saying that there are exactly two objects. Use the game to see why it is true in a world where there are two objects, but false in worlds with more or less than two objects.
4. Sentence 5 appears, at first sight, to assert that there are at least three objects, so it should be false in a world with two objects. Check to see if it is indeed false in such a world. Why isn't it? Play the game to confirm your suspicions.
5. The sixth sentence actually manages to express the claim that there are at least three objects. Do you see how it's different from the fifth sentence? Check to see that it is false in the current world, but is true if you add another object to the world.
6. The seventh sentence says that there are exactly three objects in the world. Check to see that it is true in the world with three

objects, but false if you either delete an object or add another object.

7. Sentence 8 asserts that a is a large object, and in fact the *only* large object. To see just how the sentence manages to say this, start with a world with three small objects and name one of them a. Play the game committed to true to see why the sentence is false. Now make object a large. Play the game committed to false to see why it is true. Finally, make one of the other objects large as well, and play the game committed to true to see why it is false.
8. Sentence 8 asserted that a was the only large object. How might we say that there is exactly one large object, without using a name for the object? Compare sentence 8 with sentence 9. The latter asserts that there is something which is the only large object. Check to see that it is true only in worlds in which there is exactly one large object.
9. Construct a world in which sentence 10 is true.
10. Make sentences 11 and 12 true in a single world.
11. Sentence 13 is another way to assert that there is a unique dodecahedron. That is, sentence 13 is equivalent to sentence 10. Can you see why? Check three worlds to see that the two sentences are true in the same worlds, those in which there is a single dodecahedron.
12. Sentence 14 says that there are exactly two tetrahedra. Check that it is true in such worlds, but false if there are fewer or more than two.

Problem 2. What would be good English translations of the following sentences of FOL? Figure out which of the following are logically equivalent and which are not. Explain your answers.

1. $\exists! x \mathsf{Tove}(x)$ [Remember that the notation $\exists!$ is an abbreviation, as explained above.]
2. $\exists x \forall y [\mathsf{Tove}(y) \rightarrow y = x]$
3. $\exists x \forall y [\mathsf{Tove}(y) \leftrightarrow y = x]$
4. $\forall x \forall y [(\mathsf{Tove}(x) \land \mathsf{Tove}(y)) \rightarrow x = y]$
5. $\forall x \forall y [(\mathsf{Tove}(x) \land \mathsf{Tove}(y)) \leftrightarrow x = y]$

Problem$^{\star}_{\diamond}$ 3. (Translating numerical claims) In this exercise we will try our hand at translating English sentences involving numerical claims.

- Translate the following English sentences.
  1. *There are at least two dodecahedra.*
  2. *There are at most two tetrahedra.*

3. *There are exactly two cubes.*
4. *There are only three things that are not small.*
5. *No dodecahedron is in back of the large cube.*

Save your sentences as Sentences 7.3.

- Open Peano's World. Note that all of the English sentences are true in this world. Check to see that your translations are as well.
- Open Bolzano's World. Here sentences 1, 3, and 5 are the only true ones. Verify that your translations have the right truth values in this world.
- Open Skolem's World. Only sentence 5 is true in this world. Check your translations.
- Finally, open Montague's World. In this world, sentences 2, 3, and 5 are the only true ones. Check your translations.

Problem$_\diamond$ 4. (Saying more complicated things) Open Skolem's World. Create a file called Sentences 7.4 and describe the following features of Skolem's World.

1. Use your first sentence to say that there are only cubes and tetrahedra.
2. Next say that there are exactly three cubes.
3. Express the fact that every cube has a tetrahedron that is to its right but is neither in front of or in back of it.
4. Express the fact that at least one of the tetrahedra is between two other tetrahedra.
5. Notice that the further back something is, the larger it is. Say this.
6. Note that none of the cubes is to the right of any of the other cubes. Try to say this.
7. Observe that the small tetrahedron is in front of but to neither side of all the other tetrahedra. State this.

If you have expressed yourself correctly, there is very little you can do to Skolem's World without making at least one of your sentences false. Basically, all you can do is "stretch" things out, that is, move things apart while keeping them aligned. To see this, try making the following changes.

1. Add a new tetrahedron to the world. Find one of your sentences that comes out false. Move the new tetrahedron so that a different sentence comes out false.
2. Change the size of one of the objects. What sentence now comes out false?
3. Change the shape of one of the objects. What sentence comes out false?

4. Slide one of the cubes to the left. What sentence comes out false?
5. Rearrange the three cubes. What goes wrong now?

7.2 Definite descriptions

First-order logic has only two quantifiers, whereas English has many determiners, words like "some" and "every," that combine with nouns to produce noun phrases. Other determiners include numbers (as in "two cubes") and the definite article "the" (as in "the cube"). Bertrand Russell proposed that a sentence like *The cube is small* should be analyzed as asserting that there is exactly one cube, and it is small. According to this analysis, the sentence will be false if there is no cube, or if there is more than one, or if there is exactly one, but it's not small. If this analysis is correct, then such sentences can easily be expressed in first-order logic as follows:

$$\exists x[\mathsf{Cube}(x) \land \forall y(\mathsf{Cube}(y) \to y = x) \land \mathsf{Small}(x)]$$

More generally, a sentence of the form *The P is a Q*, on the Russellian analysis, would be translated as:

$$\exists x[\mathsf{P}(x) \land \forall y(\mathsf{P}(y) \to x = y) \land \mathsf{Q}(x)]$$

We will call this the *Russellian analysis of the definite description.*

This analysis of the definite description is not without its detractors. Strawson, for example, has argued that there are uses of *the* that cannot be so translated. Suppose, for example, that someone asserts: "The elephant in my closet is not wrinkling my clothes," when in fact there is no elephant in his closet. On the Russellian analysis the claim would be false. But that would seem to suggest that the sentence "The elephant in my closet is wrinkling my clothes" is true. Strawson would want to say that in such a case *neither* of these sentences is true.

Strawson's general picture is this. Some sentences carry certain *presuppositions.* They can only be used to make a claim when those presuppositions are fulfilled. Just as you can't drive a car unless there is a car present, you cannot make a successful claim unless the presuppositions of your claim are satisfied. With our example above, the sentence could only be used to make a claim in case there was one, and only one, elephant in the speaker's closet. Otherwise the sentence simply misfires, and so does not have a truth value at all. It is much like using an FOL sentence containing a name b to describe a world where no object is named b.

If Strawson's objection is right, then there will be no general way of translating definite descriptions into FOL, since FOL sentences without names in them always have a truth value. There would be nothing to prevent one from enriching FOL to have a determiner that works this way. Indeed, this has been proposed and studied. But it does not fit into the constraints imposed by our choice of subject matter in this book.

On the other hand, there have been rejoinders to Strawson's objection. To understand them, recall our elephant in the closet. Part of the argument was that if *The elephant in my closet is not wrinkling my clothes* is false, then the sentence *The elephant in my closet is wrinkling my clothes* must be true. The rejoinder is to say that this inference is not a logical implication, but only a conversational implicature.

To see if this is plausible, we try the cancellability test. Does the following seem coherent or not? "The elephant in my closet is not wrinkling my clothes. In fact, there is no elephant in my closet." Some people think that, read with the right intonation, this makes perfectly good sense. Others disagree.

At the very least, we can say that the Russellian analysis is an important one, one that covers some uses of the English definite description. It is the one we shall be treating in the problems that follow.

Exercises and Problems

Exercise$_\diamond$ 5. (The Russellian analysis of definite descriptions)

1. Open Russell's Sentences. Sentence 1 is the second of the two ways we saw in Exercise 1 for saying that there is a single cube. Compare sentence 1 with sentence 2. Sentence 2 is the Russellian analysis of our sentence *The cube is small.* Construct a world in which sentence 2 is true.
2. Construct a world in which sentences 2-6 are all true.
3. On the Russellian analysis, sentences with two definite descriptions become remarkably complex. Sentence 7 contains the Russellian analysis of *The small dodecahedron is to the left of the medium dodecahedron.* Check to see that it is true just in those worlds in which there is a single small dodecahedron, and a single medium dodecahedron, with the former to the left of the latter.

7.3 Methods of proof involving numerical claims

Since numerical claims can all be expressed in FOL, we can use the methods of proof developed in previous chapters to prove numerical claims. However, as you may have noticed in doing the problems, numerical claims are not always terribly perspicuous when expressed in FOL notation. Indeed, expressing a numerical claim in FOL and then trying to prove the result is a recipe for disaster. It is all too easy to lose one's grasp on what needs to be proved.

Suppose, for example, you are told that there are exactly two logic classrooms, and that each classroom contains exactly three computers. Suppose you also know that every computer is in some logic classroom. From these assumptions it is of course quite easy to prove that there are exactly six computers. How would the proof go?

> **Proof:** To prove there are exactly six computers it suffices to prove that there are at least six, and at most six. To prove that there are at most six, we simply note that every computer must be in one of the two classrooms, and that each classroom contains at most three, so there can be at most six all together. To prove that there are at least six, we note that each classroom contains at least three. But now we need another assumption that was not explicitly stated in the problem. Namely, we need to know that no computer can be in two classrooms. Given that, we see that there must be at least six computers, and so exactly six.

This may seem like making pretty heavy weather of an obvious fact, but it illustrates two things. First, to prove a numerical claim of the form *there exists exactly n objects x such that P(x)*, which we agreed to abbreviate as $\exists^{!n}$xP(x), you need to prove two things: that there are *at least n* such objects, and that there are *at most n* such objects.

The proof also illustrates a point about FOL. If we were to translate our premises and desired conclusion into FOL, things would get quite complicated. And if we tried to then prove our FOL conclusion from the FOL premises using the rules we have presented earlier, we would completely lose track of the basic intuitions that make the proof work. Thus, while it would be possible to give such a proof, no one intent on giving such a proof would really write it out that way.

The problem really has to do with a syntactic shortcoming of FOL. Rather than have determiners that directly express numerical claims, they must be translated into claims expressed with ∀ and ∃. If we were to add numerical determiners to FOL, we would be able to give proofs that correspond much more closely to the intuitive proofs. Still, though, the theoretical expressive power of the language would remain the same.

We can think of the above as a new method of proof. We summarize it as follows:

Remember

When trying to prove $\exists^{!n}$xP(x), prove two things: that there are at least n objects satisfying P(x), and that there are at most n such objects.

A particularly important special case of this method is with uniqueness claims, those of the form ∃!xP(x) which say there is exactly one object with some property. To prove such a claim, we must prove two things, existence and uniqueness. In proving existence, we prove that there is at least one object satisfying P(x). Given that, we can then show uniqueness by showing that there is at most one such object. To give an example, let us prove ∃!x[Even(x) ∧ Prime(x)].

> **Proof:** We first prove existence, that is, that there is an even prime. This we do simply by noting that 2 is even and a prime. Thus, by existential generalization, there is an even prime. Next we prove uniqueness. That is, we prove that for any number x, if x is an even prime, then $x = 2$, by general conditional proof. Suppose x is an even prime. Since it is even, it must be divisible by 2. But being prime, it is divisible only by itself and 1. So $x = 2$. This concludes our proof of uniqueness.

With one significant exception (induction, which we take up in Chapter 9), we have now introduced all the major methods of proof. When these are used in mathematical proofs, it is common to suppress a lot of detail, including explicit mention of the methods being used. To a certain extent, we have already been doing this in our proofs. From now on, though, we will present proofs in a more abbreviated fashion, and expect you to be able to fill in the details. For example, here is a more abbreviated version of our proof that there is exactly one even prime. You should check to see that you could have filled in the details on your own.

> **Proof:** We first prove existence, that is, that there is an even prime. This we do simply by noting that 2 is even and a prime. We then prove uniqueness, by proving that any even prime must be 2. First, since it is even, it must be divisible by 2. But being prime, if it is divisible by 2, it is 2.

$\mathcal{F}'$: Rules for numerical quantifiers

Since the numerical quantifiers are really shorthand for more complicated expressions in our language, there is no real need to introduce rules that specifically apply to them. But it is often more convenient to have such rules, so we introduce them into our extended system $\mathcal{F}'$.

There are several ways one could state introduction and elimination rules for these quantifiers. We present two that seem especially natural and useful. We will treat only the uniqueness quantifier, ∃!, and will leave extensions to the problems that follow.

We begin with the introduction rule for this quantifier. The rule requires that you prove that a formula holds of some object and only of that object.

Uniqueness Introduction (∃! Intro):

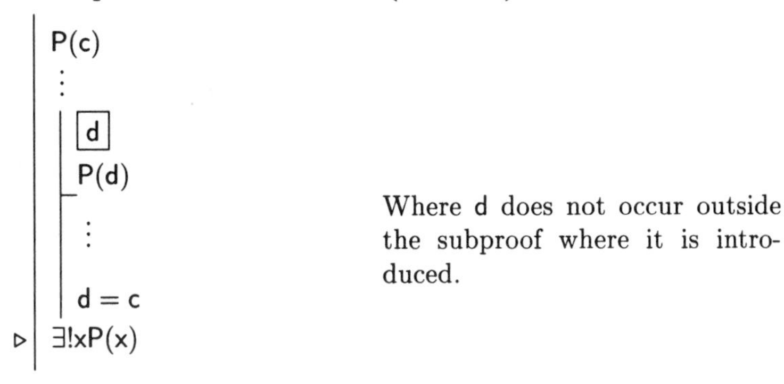

The elimination rule for the uniqueness quantifier simply allows you to extract either of the two fundamental components of the uniqueness claim. Thus there are two ways to apply the rule, corresponding to the following two schematic forms.

Uniqueness Elimination I (∃! Elim I):

$\exists!x P(x)$

$\vdots$

▷ $\exists x P(x)$

Uniqueness Elimination II (∃! Elim II):

$\exists!x P(x)$

$\vdots$

▷ $\forall y \forall z[(P(y) \land P(z)) \rightarrow y = z]$

As usual, y and z can be any two distinct variables, but we must choose them so that neither already occurs in the formula P(x).

Exercises and Problems

Problem 6. Give an informal proof that the following are logically equivalent:

1. $\exists!x \text{Tove}(x)$
2. $\exists x \text{Tove}(x) \land \forall x \forall y[(\text{Tove}(x) \land \text{Tove}(y)) \rightarrow x = y]$
3. $\exists x \forall y\, [\text{Tove}(y) \leftrightarrow y = x]$

Problem 7. Now give a formal proof in $\mathcal{F}'$ of $\exists!x \text{Tove}(x)$ from the premise $\exists x \forall y\, [\text{Tove}(y) \leftrightarrow y = x]$.

Problem 8. Give informal proofs of the following statements. The domain of discourse is the natural numbers 0, 1, ...

1. $\exists!x[x^2 - 2x + 1 = 0]$
2. $\exists^{!2}y[y + y = y \times y]$
3. $\exists^{!2}x[x^2 - 4x + 3 = 0]$
4. $\exists!x[(x^2 - 5x + 6 = 0) \wedge (x > 2)]$

Problem$_\diamond$ 9. Assume the following premises:

1. *There are exactly four cubes.*
2. *Any column that contains a cube contains a tetrahedron, and vice versa.*
3. *No tetrahedron is in back of any other tetrahedron.*

Does it follow that there are exactly four tetrahedra? If so, give an informal proof of this. If not, build a world in which all the premises are true but the conclusion is false. Save it as World 7.9.

Problem$_\diamond$ 10. Replace premise 3 of Problem 9 with the premise: *No column contains two objects of the same shape.* Does it now follow that there are exactly four tetrahedra? If so, give an informal proof of this. If not, build a world in which all the premises are true but the conclusion is false. Save it as World 7.10.

Problem* 11. Give introduction and elimination rules for the quantifier $\exists^{!2}$.

7.4 Some review problems

In this section we present some moderately challenging problems intended to review various topics discussed in the main sections of this book. The problems also introduce some topics we have not dealt with in the book.

Problem$_\diamond$ 12. (Games of incomplete information) As you recall, you can sometimes know that a sentence is true in a world without knowing how to play the game and win. Open Mostowski's World. Translate the following into first-order logic. Save your sentences as Sentences 7.12. Now, without using the 2-D view, make as good a guess as you can about whether the sentences are true or not in the world. Once you have assessed a given sentence, use **Verify** to see if you are right. Then, with the correct truth value checked, see how far you can go in playing the game. Quit whenever you get stuck, and play again. Can you predict in advance when you will be able to win? Do not look at the 2-D view until you have finished the whole exercise.

1. *There are at least two tetrahedra.*
2. *There are at least three tetrahedra.*
3. *There are at least two dodecahedra.*
4. *There are at least three dodecahedra.*
5. *Either there is a small tetrahedron behind a small cube or there isn't.*
6. *Every large cube is in front of something.*
7. *Every tetrahedron is in back of something.*
8. *Every small cube is in back of something.*
9. *Every cube has something behind it.*
10. *Every dodecahedron is small, medium, or large.*
11. *If e is to the left of every dodecahedron, then it is not a dodecahedron.*

Now modify the world so that the true sentences are still true, but so that it will be clear how to play the game and win. Save the world as World 7.12.

Problem$^{\star}_{\diamond}$ 13. (Validity$_I$ *versus* Validity$_U$) As we have pointed out, there is one significant difference between validity (logical truth) for a language like that used in Tarski's World and validity for the kinds of languages usually studied in logic text books. The difference stems from the fact that there are logical relations among the atomic sentences of Tarski's World. Most logic books assume that the basic relation symbols of the language, other than =, are "uninterpreted." Thus, for example, they would assume that there was no logical relationship between the atomic sentences LeftOf(a, b) and RightOf(b, a). As a result, they would end up having a slightly different notion of validity.

In order to talk coherently about the two notions and compare them, let's use a subscript I for the notion of validity applied to an interpreted language, and U for the notion applied when the predicate symbols (other than =) are treated as uninterpreted. Anything that is valid$_U$ is valid$_I$, but in general the converse does not hold. Thus, for example, the sentence

$$\forall x\, \forall y\, (\mathsf{LeftOf}(x, y) \rightarrow \mathsf{RightOf}(y, x))$$

is valid$_I$ but not valid$_U$. If a sentence is valid$_I$ but not valid$_U$, then there must be a way to reinterpret the predicate symbols so that the result can be falsified in some world. Open Carnap's Sentences and Bolzano's World.

1. Paraphrase each sentence in English and verify that it is true in the given world.
2. For each sentence, decide whether you think it is true in *all* worlds or not, that is, whether it is valid$_I$ or not. If it is not valid$_I$,

find a world in which the the sentence comes out false, and save the world as World 7.13.x, where x is the number of the sentence. [Hint: Exactly five of them are valid$_I$.]

3. Which of these sentences are valid$_U$? [Hint: Only one is.]
4. For each sentence which is valid$_I$ but not valid$_U$, think of a way to reinterpret the predicates in the sentence so that the result can be falsified in some world. Feel free to specify interpretations that cannot be expressed in the current language, e.g., *x is 5 squares away from y.* Of course, your interpretations must respect the arity of the original predicates.

Problem$^{\star\star}_{\diamond}$ 14. (Validity$_I$ *versus* non-logical truth in all worlds) Another distinction Tarski's World helps us to understand is the difference between sentences that are valid$_I$ and sentences that are, for reasons that have nothing to do with logic, true in all worlds. The notion of valid$_I$ has to do with a sentence being true simply in virtue of the *meaning* of the sentence, and so no matter how the world is. However, some sentences are true no matter how the world is, not because of the meaning of the sentence or its parts, but because of, say, laws governing the world. We can think of the constraints imposed by the innards of Tarski's World as physical laws governing how the world can be. For example, the sentence which asserts that there are at most 12 objects happens to hold in all the worlds that we can construct with Tarski's World. However, it is not valid$_I$, let alone valid$_U$.

Open Post's Sentences. Classify each sentence in one of the following ways: (A) valid$_I$, (B) true in all worlds that can be depicted using Tarski's World, but not valid$_I$, or (C) falsifiable in some world that can be depicted by Tarski's World. For each sentence of type (C), build a world in which it is false, and save it as World 7.14.x, where x is the number of the sentence. For each sentence of type (B), use a pencil and paper to depict a world in which it is false. (In doing this exercise, assume that Medium simply means *neither small nor large*, which seems plausible. However, it is not plausible to assume that Cube means *neither a dodecahedron nor tetrahedron*, so you should not assume anything like this.)

Problem$_{\diamond}$ 15. (Satisfiability) Recall that a sentence is satisfiable if there is possible circumstance in which it is true. Like validity, the notion of satisfiability can be divided into *satisfiable$_I$* and *satisfiable$_U$*, depending on whether the meanings of the predicates are assumed to be fixed. This problem is about satisfiability$_I$. Determine whether the following set of sentences is satisfiable. If it is, build a world in which all the sentences are true. Save it as world World 7.15. If it is not, derive a contradiction from the set.

1. *Every cube is to the left of every tetrahedron.*
2. *There are no dodecahedra.*
3. *There are exactly four cubes.*
4. *There are exactly four tetrahedra.*
5. *No tetrahedron is large.*
6. *Nothing is larger than anything to its right.*
7. *One thing is to the left of another just in case the latter is behind the former.*

Problem$^{\star}_{\diamond}$ 16. (More about satisfiability) Open Padoa's Sentences. Here you will find a set of four sentences.

1. Any three of these sentences form a satisfiable set. There are four sets of three sentences, so to show this, build four worlds, World 7.16.123, World 7.16.124, World 7.16.134, and World 7.16.234, where the four sets are true. (Thus, for example, sentences 1, 2 and 4 should be true in World 7.16.124.)
2. Explain why you can't build a world that satisfies all four sentences. Is the set of sentences unsatisfiable$_I$? [Hint: Imagine a world where one of the blocks is a sphere.]
3. Reinterpret the predicates Tet and Dodec in such a way that sentence 3 comes out true in World 7.16.124. Since this is the only sentence that uses these predicates, it follows that all four sentences would, with this reinterpretation, be true in this world. (This shows that the set is satisfiable$_U$.)
4. Reinterpret the predicate Between in such a way that World 7.16.123 makes all the sentences true.

Problem$_{\diamond}$ 17. (Vacuous quantifiers)

1. Open Finsler's Sentences and Gödel's World. Verify that each of the sentences is true in the world. Six of these sentences contain one or more vacuous quantifiers. See if you spot them.
2. Now play the game on each of the sentences you think has a vacuous quantifier, committed to its truth. What happens when Tarski's World encounters a vacuous quantifier?
3. At the end of this list of sentences, rewrite the sentences you think have vacuous quantifiers, so that there is no vacuous quantification. Check that your answers are indeed true sentences.
4. How can you describe the semantic function of vacuous quantification?
5. Look at sentence 8. Neither of the quantifiers in this sentence is vacuous, however the second quantifier binds the same variable as the first. Figure out what this sentence says by assessing its truth

and playing the game. Write a sentence at the end equivalent to it, but one that uses two different variables.

6. Do you think there could ever be a need to append a quantifier $\forall$y or $\exists$y to a formula that already contains the variable y bound by another quantifier (as in sentence 8)? Before answering, you might want to do the next two exercises.

Save the enlarged sentence file as Sentences 7.17.

Problem$^{\star\star}_{\diamond}$ 18. (Numbers of variables) Tarski's World only allows you to use six variables. Let's explore what kind of limitation this imposes on our language.

1. Translate the sentence *There are at least two objects*, using only the predicate =. How many variables do you need?
2. Translate *There are at least three objects.* How many variables do you need?
3. Can you prove that you cannot express the sentence *There are at least seven objects* using only = and the six variables available in Tarski's World? [Warning: This is true, but it is not easy to prove.]

Save the sentences used for the first two in Sentences 7.18.

Problem$_{\diamond}$ 19. (Reusing variables) In spite of the above exercise, there are in fact sentences we can express using just the six available variables that can only be true in worlds with at least seven objects. For example, in Robinson's Sentences, we give such a sentence, one that only uses the variables x and y.

1. Open this file. Click on the cube button six times and test the sentence's truth. Now add one more small cube, and test the sentence's truth again. Then play the game committed (incorrectly) to false. Can you see the pattern in Tarski's World's choice of objects? When it needs to pick an object for the variable x, it picks the leftmost object to the right of all the previous choices. Then, when it needs to pick an object for the variable y, it picks the last object chosen. Can you now see how the reused variables are working?
2. Now delete one of the cubes, and play the game committed (incorrectly) to true. Do you see why you can't win?
3. Now write a sentence which says there are at least four objects, one in front of the next. Use only variables x and y. Save this sentence in Sentences 7.19.

Problem** 20. (Persistence through expansion) As we saw in Exercise 4, page 156, some sentences simply can't be made false by adding objects of various sorts to the world. Once they are true, they stay true. For example, the sentence *There is at least one cube and one tetrahedron*, if true, cannot be made false by adding objects to the world. This exercise delves into the analysis of this phenomenon in a bit more depth.

Let's say that a sentence A is *persistent through expansion* if, whenever it is true, it remains true no matter how many objects are added to the world. (In logic books, this is usually called just persistence, or persistence under extensions.) Notice that this is a semantic notion. That is, it's defined in terms of truth in worlds. But there is a corresponding syntactic notion. Call a sentence *existential* if it is logically equivalent to a prenex sentence containing only existential quantifiers.

- Show that Cube(a) → ∃x FrontOf(x, a) is an existential sentence.
- Is ∃x FrontOf(x, a) → Cube(a) an existential sentence?
- Show that every existential sentence is persistent through expansion. [Hint: You will have to prove something slightly stronger, by induction on wffs. If you are not familiar with induction on wffs, just try to understand why this is the case. If you are familiar with induction, try to give a rigorous proof.] Conclude that every sentence equivalent to an existential sentence is persistent through expansion.

It is a theorem, due to Tarski and Łoś (a Polish logician whose name is pronounced more like "wash" than like "loss"), that any sentence which is persistent through expansion is existential. Since this is the converse of what you were asked to prove, we can conclude that a sentence is persistent through expansion if and only if it is existential. This is a classic example of a theorem that gives a syntactic characterization of some semantic notion. For a proof of the theorem, see any textbook in model theory.

Problem$_\diamond$ 21. (Invariance under motion, part 1) The real world does not hold still, the way the world of mathematical objects does. Things move around. The truth values of some sentences change with such motion, while the truth values of other sentences don't. Open Ockham's World and Ockham's Sentences. Verify that all the sentences are true in the given world. Make as many of Ockham's Sentences false as you can by just moving objects around. Don't add or remove any objects from the world, or change their size or shape. You should be able to make false (in a single world) all of the sentences containing any spatial predicates, that is, containing LeftOf, RightOf, FrontOf, BackOf, or Between. (However, this is a quirk of this list of sentences, as we will see in the next exercise.) Save the world as World 7.21.

Problem 22.** (Invariance under motion, part 2) Call a sentence *invariant under motion* if, for every world, the truth value of the sentence (whether true *or* false) does not vary as objects move around in that world.

1. Prove that if a sentence does not contain any spatial predicates, then it is invariant under motion.
2. Give an example of a sentence containing a spatial predicate that is nonetheless invariant under motion.
3. Give another such example. But this time, make sure your sentence is not logically equivalent to any sentence that doesn't contain spatial predicates.

Problem◇ 23. (Persistence under growth, part 1) In the real world, things not only move around, they also grow larger. (Some things also shrink, but ignore that for now.) Starting with Ockham's World, make the following sentences true by allowing some of the objects to grow:

1. ∀x ¬Small(x)
2. ∃x ∃y (Cube(x) ∧ Dodec(y) ∧ Larger(y, x))
3. ∀y (Cube(y) → ∀v (v ≠ y → Larger(v, y)))
4. ¬∃x∃y(¬Large(x) ∧ ¬Large(y) ∧ x ≠ y)

How many of Ockham's Sentences are false in this world? Save your world as World 7.23.

Problem* 24. (Persistence under growth, part 2) Say that a sentence A is *persistent under growth* if, for every world in which A is true, A remains true if some or all of the objects in that world get larger. Thus, Large(a) and ¬Small(a) are persistent under growth, but Smaller(a,b) isn't. Give a syntactic definition of as large a set of sentences as you can for which every sentence in the set is persistent under growth. Can you prove that all of these sentences are persistent under growth?

Problem*◇ 25. (Inexpressibility) Let's say that two worlds W_1 and W_2 are *elementarily equivalent* for our language just in case exactly the same sentences of the language are true in both worlds. This notion can be applied using any first-order language, and with any two worlds. Sometimes worlds will be elementarily equivalent for one language, but not for another.

Worlds can be quite different but still be elementarily equivalent with respect to some language. This happens when there are differences in the worlds that simply cannot be captured by any sentence of the given language. This is a very important notion in first-order logic (one that was introduced, by the way, by Tarski). We can illustrate it quite simply with Tarski's World.

- Build a world in which the following sentences are true:
  1. *There are exactly three objects, a cube, a dodecahedron, and a tetrahedron.*
  2. *Every object is small.*
  3. *The cube is to the left of the dodecahedron.*
  4. *Something is between the cube and the dodecahedron.*
  5. *The cube is immediately next to the tetrahedron, which is immediately next to the dodecahedron.*

  Save this world as World 7.25.
- Using our first-order language, give as faithful a rendering as possible of these English sentences. Check that they are true in your world. Save your sentences as Sentences 7.25.
- Now alter the world by sliding the three objects apart, so that they are no longer immediately next to one another. Show that all the translations you came up with are still true. Thus, your translation of the last sentence must not have captured what was said by the English sentence.
- (⋆⋆⋆) Prove that, in fact, the two worlds are elementarily equivalent.

This shows that the relation of one object being immediately next to another simply cannot be expressed in the language we've used in Tarski's World, no matter how clever or complicated a sentence we might come up with. Of course one could consider a richer language, where the predicate Adjoins was included. But the phenomenon of essentially distinct, but elementarily equivalent worlds is almost always present in first-order languages.

7.5 Expressive limitations of first-order logic

By now you have learned how to translate quite complex claims from English into FOL. However, we should not close this chapter without pointing out that there are many expressions of natural languages like English which go beyond first-order logic in various ways.

Some of these we have already discussed at various points, both with examples and problems. As one example, we saw that there are many uses of the natural language conditional *if* ... *then* ... that are not truth-functional, and so not captured by the truth-functional connective $\rightarrow$.

Another disparity between English and FOL comes in the range of determiners available in the former, and the limited number of quantifiers in the latter. In this chapter, we have seen how some of the latter can be captured by convoluted circumlocutions, especially the numerical determiners, as in *Two cubes are large.* But there are many determiners that are not expressible in FOL. Take the determiner *Most*, as in *Most cubes are large.* There are two problems, here. One is that the meaning of *most* is a bit vague. *Most cubes are large* clearly implies *More than half the cubes are large*, but what about the converse? Intuitions differ. But even if we take it to mean the same as *More than half*, it cannot be expressed in FOL. Other examples of useful English determiners that go beyond the expressive power of FOL are expressions like *many*, *few*, *lots of the*, and so forth. These expressions do not lend themselves to translation into FOL.

A related difference in expressive power comes in the ability of English to use both singular and plural noun phrases. There is a difference between saying *Any student could get an A in this course* and saying *All students could get an A in this course.* If the instructor grades on a curve, and all students are of equal ability, the first is probably true but the second false. FOL does not allow us to capture this difference.

Another dimension in which FOL is limited, as contrasted with natural languages, comes in the flexible use of tense. FOL assumes a timeless domain of unchanging relationships, whereas in English, we can exploit our location in time and space to say things about the present, the past, and locations around us. For example, in FOL we cannot say that it is hot here today but it was cool yesterday. To say something similar in FOL, we would need to allow quantifiers over times and locations, and add corresponding new argument positions to our atomic predicates.

Similarly, languages like English have a rich modal structure, allowing us not just to say how things are, but how they must be, how they might be, how they can't be, how they should be, how they would be if we had our way, and so forth. So, for example, we can say *All the king's horses couldn't put Humpty Dumpty together again.* Or *Humpty shouldn't have climbed on the wall.* Or *Humpty might be dead.* Such statements lie outside the realm of FOL.

All of these expressions have their own logic, in that we can explore and try to understand just what sort of claims involving these expressions follow from others. Building on the successes of FOL, logicians have studied (and are continuing to study) extensions of FOL in which these and similar expressive deficiencies are addressed. But as of now there is no single such extension of FOL that has gained anything like its currency.

Exercises and Problems

Exercise 26. Try to translate the nursery rhyme about Humpty Dumpty into FOL. Point out the various linguistic mechanisms that go beyond FOL. Discuss this in class.

Problem 27. Is the following argument logically valid? If so, explain why. If not, describe a situation in which the premises are true but the conclusion false. Take the domain of discourse to be Claire's and Max's kindergarten class.

1. *Most of the students brought a snack.*
2. *Most of the students got here late.*
3. Conclusion: *Most of the students got here late and brought a snack.*

Problem 28. Is the following a logical consequence of the premises in Problem 27?

Some student got here late and brought a snack.

If so, explain why. If not, describe a situation in which the premises are true but the conclusion false.

Problem 29. Does it follow from the fact that *Few cubes are large* that *Few cubes are large cubes*? If so, explain why. If not, describe a situation in which the premises are true but the conclusion false.

Problem 30. Does it follow from the fact that *Few cubes are large* that *Few large things are cubes*? If so, explain why. If not, describe a situation in which the premises are true but the conclusion false.

Problem 31. Consider the following two claims. Does either follow logically from the other? Are they logically equivalent? Explain your answers.

1. I can eat every apple in the bowl.
2. I can eat any apple in the bowl.

Problem* 32. Recall the first-order language introduced in Table 1, page 23. Some of the following can be given first-order translations using that language, some cannot. Translate those that can be. For the others, explain why they cannot be faithfully translated, and discuss whether they could be translated with additional names, predicates, and function symbols, or if the shortcoming in the language is more serious.

1. *Claire gave Max at least two disks at 2:00 pm.*
2. *Claire gave Max at most two disks at 2:00 pm.*
3. *Claire gave Max several disks at 2:00 pm.*
4. *Claire was a student before Max was.*
5. *The disk Max gave Claire at 2:00 pm was blank.*
6. *Most disks were blank at noon.*
7. *All but two disks were blank at noon.*
8. *There is at least one student who made Max angry every time he (or she) gave Max a disk.*
9. *Max was angry whenever a particular student gave him a disk.*
10. *If someone gave Max a disk, it must have been Claire.*

11. *No disk erased by Max between 2:00 and 2:05 belonged to Claire.*
12. *If Claire erased one of Max's disks before 2:00 pm, then Max was angry at 2:00 pm.*
13. *Folly's owner was a student.*
14. *Before 3:00, no one gave anyone a disk unless it was blank.*
15. *No one should give anyone a disk unless it is blank.*
16. *A disk that is not blank always belongs to someone or other.*
17. *A disk that is not blank must belong to someone or other.*
18. *Max was angry at 2:00 pm because Claire had erased one of his disks.*
19. *When Max gave Folly to Claire, Folly was blank, but it was not blank five minutes later.*
20. *No student could possibly be a disk.*

Problem 33.** Here is a famous puzzle. There was a Roman who went by two names, "Cicero" and "Tully." Discuss the validity or invalidity of the following argument.

1. Bill claims Cicero was a great orator.
2. Cicero is Tully.
3. Bill claims Tully was a great orator.

What is at stake here is nothing more or less than the principle that if $(\ldots\ a \ldots)$ is true, and $a = b$, then $(\ldots\ b \ldots)$ is true. [The puzzle is usually stated with "believes" rather than "claims."]

Part III

Applications of First-order Logic

8

First-order Set Theory

Over the past hundred years, set theory has come to be a very useful part of mathematics. It is used both in mathematics itself, as a sort of universal language, and also in applications outside of mathematics, especially in the symbolic sciences. The reason set theory is so useful is that sets provide us with a medium we can use to model many other kinds of structures. In this chapter we apply first-order logic to the study of set theory. Since the chapter is rather unusual in its presentation, we start with an outline, by way of warning.

We study the intuitive notion of set, the one that you have no doubt been exposed to in some form or other in school. We begin by isolating two basic principles that seem, at first sight, to be clearly true of this intuitive notion of set. The principles are called the Axiom of Extensionality and the Axiom of Comprehension.[1] We state the axioms in the first-order language of set theory.

Having expressed the axioms in FOL, we are in the position of being able to apply all that we have learned earlier in this book. For example, we can study the logical consequences of our axioms, and use the methods of proof we have studied to draw out these consequences. This provides an excellent testing ground for you to see if you can use these methods in doing some real mathematics. Of course, in a single chapter we will only be able to scratch the surface of the subject.

We don't have to go far, however, before we discover that we can prove a contradiction from the axioms. This contradiction will demonstrate that the axioms are inconsistent. There cannot really be a domain of discourse satisfying our axioms: the intuitive notion of a set is just inconsistent. The inconsistency boils down to a simple set-theoretic paradox, known as Russell's Paradox.

[1] Axioms are the same as premises, with the intuitive difference that axioms are usually more permanent and widely believed. "Logic is the last refuge of scoundrels," might be a premise, but surely no one's axiom. "I don't trust real estate agents" is an axiom.

Russell's Paradox has had a profound impact on modern set theory and logic. It forced the founders of set theory to go back and think more critically about the intuitive notion of a set. The aim was to refine it in a way that avoids inconsistency. Examining this refined version of the notion, forces us to modify one of our axioms, in order to have a consistent theory of sets. We end the chapter by stating the revised axioms that make up modern set theory.

This may seem like a rather tortured route to the modern theory, but it is very hard to understand and appreciate modern set theory without first being exposed to the original version, understanding what is wrong with it, and seeing how the modern theory derives from this understanding.

8.1 Cantor's set theory

We will concentrate on the intuitive conception of sets first studied extensively by Georg Cantor in the nineteenth century. Intuitively, a set is just a collection of things, like a set of chairs, a set of dominos, or a set of numbers. The things in the collection are said to be *members* of the set. We write $a \in b$, and read "a is a member (or an element) of b," if a is one of the objects that makes up the set b. If we can list all the members of b, say the numbers 7, 8, and 10, then we write $b = \{7, 8, 10\}$. This is called a "list description" of the set.

As you will recall, the first-order language of set theory has two relation symbols, $=$ and $\in$. Beyond that, there are some options. If we want our domain of discourse to include not just sets but other things as well, then we need some way to distinguish sets from other objects. One way that is sometimes used is to have a unary predicate symbol $Set(x)$ that picks out the sets. An alternate way of accomplishing the same thing, one that results in formulas that are easier to read, is to have two sorts of variables. One sort has a domain that includes all and only the sets, while the other sort ranges over the sets and any other objects in the domain of discourse. This is the approach we will take. We use variables a, b, c, with and without subscripts, to range over sets, and variables $x, y, z, \ldots$ to range over everything. Thus, for example, if we wanted to say that everything is a member of some set or other, we would write

$$\forall x \exists a (x \in a)$$

To say the same thing with one sort of variable and the predicate $Set(x)$, we would have to write

$$\forall x \exists y [Set(y) \wedge x \in y]$$

There are other symbols that get used a lot in doing set theory. For example, there is an individual constant $\emptyset$ used to denote the empty set, a relation symbol $\subseteq$ used to express the subset relation, and function symbols $\cap$ (intersection) and $\cup$ (union). We will discuss the status of these other symbols later on.

The Axiom of Extensionality

There are two principles that capture Cantor's conception of a set. One is that a set is completely determined by its members. If you know the members of a set b, then you know everything there is to know about the identity of the set. This principle is captured by the *Axiom of Extensionality* of sets. Stated more precisely, the Axiom of Extensionality claims that if sets a and b have the same elements, then $a = b$.

We can express this in FOL as follows:

$$\forall a \forall b[\forall x(x \in a \leftrightarrow x \in b) \rightarrow a = b]$$

In particular, the identity of a set does not depend on how it is described. For example, the set of prime numbers between 6 and 12 is the very same set as the set of solutions to the equation $x^2 - 18x + 77 = 0$. This set can also be described as $\{7, 11\}$, or as $\{11, 7\}$, or in countless other ways.

Notice that if we were developing a theory of properties, we would not take this as an axiom. It is perfectly reasonable to have two distinct properties which apply to exactly the same thing. For example, the property of being a prime number between 6 and 12 is a different property than that of being a solution to the equation $x^2 - 18x + 77 = 0$, even though it happens that the property holds of the same individuals.

The Axiom of Comprehension

The second principle behind Cantor's set theory is the so-called *Unrestricted Comprehension Axiom.* It states, roughly, that every property determines a set. That is, given any determinate property P, there is a set of all objects that have this property. Thus, for example, there is a set of objects that have the following property: being an integer greater than 6 and less than 10. This is the set usually written as $\{7, 8, 9\}$.

This way of stating the Axiom of Comprehension has a certain problem, namely it talks about properties. We don't want to get into the business of having to axiomatize properties as well as sets. To get around this, we use formulas of first-order logic. Thus, for each formula $P(x)$ of the language of set theory, we take as a basic axiom the following:

$$\exists a \forall x[x \in a \leftrightarrow P(x)]$$

This says that there is a set a whose members are all and only those things that satisfy $P(x)$. (To make sure it says this, we demand that the variable a not be free in the wff $P(x)$.)

Notice that this is not just one axiom, but an infinite collection of axioms, one for each wff $P(x)$. For this reason, it is called an "axiom scheme." We will see later that some instances of this axiom scheme are inconsistent, so we will have to modify the scheme. But for now we assume all of its instances as axioms in our theory of sets.

Actually, the Axiom of Comprehension is a bit more general than our notation suggests, since the wff $P(x)$ can contain variables other than x, say $z_1, \ldots, z_n$. What we really want is the "universal closure" of the displayed formula, where all the other variables are universally quantified:

$$\forall z_1 \ldots \forall z_n \exists a \forall x[x \in a \leftrightarrow P(x)]$$

Most applications of the axiom will in fact make use of other variables.

In some ways, the Axiom of Comprehension, as we have stated it, is weaker than the intuitive principle that motivated it. After all, we have already seen that there are many determinate properties expressible in English that cannot be expressed in FOL. These are getting left out of our axiomatization. Still, the axiom as stated is quite strong. In fact, it is too strong, as we will see.

Combining the Axioms of Extensionality and Comprehension allows us to strengthen the Axiom of Comprehension. Namely, for each wff $P(x)$ we can prove that there is a *unique* set of objects that satisfy $P(x)$. In first-order notation:

$$\forall z_1 \ldots \forall z_n \exists ! a \forall x[x \in a \leftrightarrow P(x)]$$

Proof: Let $z_1, \ldots, z_n$ be arbitrary objects. The Axiom of Comprehension assures us that there is at least one set of objects that satisfy $P(x)$. So we need only prove that there is at most one. Suppose a and b are both sets that have as members exactly those things that satisfy $P(x)$. That is, a and b satisfy:

$$\forall x[x \in a \leftrightarrow P(x)]$$
$$\forall x[x \in b \leftrightarrow P(x)]$$

But then it follows that a and b satisfy:

$$\forall x[x \in a \leftrightarrow x \in b]$$

(This actually uses a variety of the methods of proof we have discussed. Could you write it out in complete detail?) So, by the Axiom of Extensionality, $a = b$.

This shows that given any first-order wff $P(x)$, our axioms allow us to deduce the existence of *the* set of objects that satisfy that wff. The set of all objects x that satisfy $P(x)$ is often written informally as follows:

$$\{x \mid P(x)\}$$

This is read: "the set of x such that $P(x)$." Note that if we had used a different variable, say "y" rather than "x," we would have had different notation for the very same set:

$$\{y \mid P(y)\}$$

This "brace notation" for sets is, like list notation, convenient but inessential. Neither is part of the official first-order language of set theory, since they do not fit into the format of first-order languages. We will only use them in informal contexts.

Exercises and Problems

Problem 1. Are the following true or false? Prove your claims.

1. $\{7, 8, 9\} = \{7, 8, 10\}$
2. $\{7, 8, 9, 10\} = \{7, 8, 10, 9\}$
3. $\{7, 8, 9, 9\} = \{7, 8, 9\}$

Problem 2. Give a list description of the following sets.

1. The set of all prime numbers between 5 and 15.
2. $\{ x \mid x$ is a member of your family $\}$
3. The set of letters of the English alphabet.
4. The set of words of English with three successive double letters.

Problem 3. List three members of the sets defined by the following properties:

1. Being a prime number larger than 15.
2. Being one of your ancestors.
3. Being a grammatical sentence of English.
4. Being a prefix of English.
5. Being a palindrome of English, that is, a phrase whose reverse is the very same phrase, as with "Madam, I'm Adam".

Problem 4. Are the following true or false?

1. $y \in \{x \mid x$ is a prime less than $10\}$ if and only if y is one of 2, 3, 5 and 7.
2. $\{x \mid x$ is a prime less than $10\} = \{2, 3, 5, 7\}$.
3. Ronald Reagan $\in \{x \mid x$ was President of the US$\}$.
4. "Ronald Reagan" $\in \{x \mid x$ was President of the US$\}$.

8.2 Singletons and the empty set

There are two special kinds of sets that sometimes cause confusion. One is where the set is obtained using the Axiom of Comprehension, but with a property that is satisfied by exactly one object. The other is when there is no object at all satisfying the property. Let's take these up in turn.

Suppose there is one and only one object x satisfying $P(x)$. According to the Axiom of Comprehension, there is a set, call it a, whose only member is x. That is, $a = \{x\}$. Some students are tempted to think that $a = x$. But in that direction lies, if not madness, at least dreadful confusion. After all, a is a set (an abstract object) and x might have been any object at all, say Stanford's Hoover Tower. Hoover Tower is a physical object, not a set. So we must not confuse an object x with the set $\{x\}$, called the *singleton set containing* x. Even if x is a set, we must not confuse it with its own singleton. For example, x might have any number of elements in it, but $\{x\}$ has exactly one element: x.

The other slightly confusing case is when nothing at all satisfies $P(x)$. Suppose, for example, that $P(x)$ is the formula $x \neq x$. In this case what are we to make of the set

$$\{x \mid x \neq x\}?$$

Well, in this case the set is said to be *empty*, since it contains nothing. It is easy to prove that there can be at most one such set, so it is called *the* empty set, and is denoted by $\emptyset$. Some authors use 0 to denote the empty set. It can also be informally denoted by $\{\}$.

Exercises and Problems

Problem 5. Suppose that a_1 and a_2 are sets, each of which has only Hoover Tower as a member. Show that $a_1 = a_2$.

Problem 6. Prove that there is only one empty set. (Hint: Use the Axiom of Extensionality.)

Problem 7. Prove that the set of even primes greater than 10 is equal to the set of even primes greater than 100.

8.3 The subset relation

Here is an important definition.

Definition 1. Given sets a and b, we say that a is a *subset* of b, written $a \subseteq b$, provided every member of a is also a member of b.

Thus, for example, the set of vowels, $\{a, e, i, \ldots\}$, is a subset of the set of letters of the alphabet, $\{a, b, c, \ldots\}$, but not vice versa.

It is very important to read the sentences "$a \in b$" and "$a \subseteq b$" carefully. The first is read "a is a member of b" or "a is an element of b." The latter is read "a is a subset of b." Sometimes it is tempting to read one or the other of these as "a is included in b". However, this is a very bad idea, since the term "included" is ambiguous between membership and subset. (If you insist on using "included," use it only for subset.)

From the point of view of FOL, there are two ways to think of Definition 1. One is to think of it as saying that the formula "$a \subseteq b$" is an abbreviation of the following wff:

$$\forall x[x \in a \rightarrow x \in b]$$

Another way is to think of $\subseteq$ as an additional binary relation symbol in our language, and to construe the definition as an axiom:

$$\forall a \forall b[a \subseteq b \leftrightarrow \forall x(x \in a \rightarrow x \in b)]$$

It does not make much difference which way one thinks of it. Different people prefer different understandings. The first is probably the most common, since it keeps the official language of set theory pretty sparse.

The following theorem is very easy to prove, but it is also extremely useful. You will have many opportunities to use it in what follows.

Theorem 1. *Given sets a and b, $a = b$ if and only if $a \subseteq b$ and $b \subseteq a$. In symbols:*

$$\forall a \forall b(a = b \leftrightarrow a \subseteq b \land b \subseteq a)$$

Proof: Let a and b be arbitrary sets. We first prove that if $a = b$ then $a \subseteq b$ and $b \subseteq a$. So, assume that $a = b$. We need to prove that $a \subseteq b$ and $b \subseteq a$. But if a and b are the very same set, then certainly every member of a is a member of b and vice versa.

Next we assume that $a \subseteq b$ and $b \subseteq a$ and prove that $a = b$. To prove this, we use the Axiom of Extensionality. By that axiom, it suffices to prove that a and b have the same members. But this follows from our assumption, which amounts to the assumption that every member of a is a member of b and vice versa.

Since a and b were arbitrary sets, our proof is complete.

Remember

Let a and b be sets.

1. $a \subseteq b$ iff every element of a is an element of b.
2. $a = b$ iff $a \subseteq b$ and $b \subseteq a$.

Exercises and Problems

Problem 8. Prove the following simple theorem: *For every set a, $\emptyset \subseteq a$.*

Problem 9. Which of the following are true?

1. The set of all US senators $\subseteq$ the set of US citizens.
2. The set of all Stanford students $\subseteq$ the set of US citizens.

3. The set of all male Stanford students $\subseteq$ the set of all males.
4. The set of all John's brothers is $\subseteq$ the set of all John's relatives.
5. The set of all John's relatives $\subseteq$ the set of all John's brothers.
6. $\{2,3,4\} \subseteq \{1+1, 1+2, 1+3, 1+4\}$
7. $\{\text{“2”}, \text{“3”}, \text{“4”}\} \subseteq \{\text{“1 + 1”}, \text{“1 + 2”}, \text{“1 + 3”}, \text{“1 + 4”}\}$

8.4 Intersection and union

Again, we start off with a definition.

Definition 2. Let a and b be any two sets.

1. The *intersection* of a and b is the set whose members are just those objects in both a and b. This set is written: $a \cap b$.
2. The *union* of a and b is the set whose members are just those objects in either a or b or both. This set is written: $a \cup b$.

At first sight, these definitions seem no more problematic than the definition of the subset relation. But if you think about it, you will see that there is actually something a bit fishy about them, as they stand. For how we do know that there *are* sets of the kind described. For example, how do we know that there is a set whose members are the objects in both a and b? And how do we know that there is exactly one such set? Remember the rules of the road. We have to prove everything from explicitly given axioms. Can we prove that there is such a unique set with our axioms?

It turns out that we can. When we later modify the Axiom of Comprehension, however, the first of these will be provable and the second won't. But since both are intuitively true, the one that is not provable from the modified axiom will then be thrown in as a new axiom.

Theorem 2. (Intersection) *For any pair of sets a and b there is one and only one set c whose members are the objects in both a and b. In symbols:*

$$\forall a \forall b \exists ! c \forall x (x \in c \leftrightarrow (x \in a \land x \in b))$$

Proof: We use general conditional proof. Thus, let a and b be arbitrary sets. We must prove that there is such a set c, and that there is only one such set. To prove existence, we use the Axiom of Comprehension. Let $c = \{x \mid x \in a \land x \in b\}$. This set clearly has the desired members. But the Axiom of Extensionality assures us that there is only one such set.

Theorem 3. (Union) *For any pair of sets a and b there is one and only one set c whose members are the objects in either a or b or both. In symbols:*

$$\forall a \forall b \exists ! c \forall x (x \in c \leftrightarrow (x \in a \lor x \in b))$$

Proof: Again, let a and b be arbitrary sets. We need to prove that there is such a set c, and that there is only one such set. To prove that there is at least one, we again use the Axiom of Comprehension. Let $c = \{x \mid x \in a \vee x \in b\}$. This set clearly has the desired members. Again the Axiom of Extensionality assures us that there is only one such set.

Here are a number of theorems which we can prove using the above definitions and results.

Theorem 4. *Let a, b, and c be any sets.*

1. $a \cap b = b \cap a$
2. $a \cup b = b \cup a$
3. $a \cap b = b$ *if and only if* $b \subseteq a$
4. $a \cup b = b$ *if and only if* $a \subseteq b$
5. $a \cap (b \cup c) = (a \cap b) \cup (a \cap c)$
6. $a \cup (b \cap c) = (a \cup b) \cap (a \cup c)$

Proof: We prove one of these and leave the rest as problems. Since (3) is the most interesting, we prove it. Let a and b be arbitrary sets. We need to prove $a \cap b = b$ iff $b \subseteq a$. To prove this, we give two conditional proofs. First, assume $a \cap b = b$. We need to prove that $b \subseteq a$. But this means $\forall x[x \in b \rightarrow x \in a]$. So, we will use the method of general conditional proof. Let x be an arbitrary member of b. We need to show that $x \in a$. But since $b = a \cap b$, we see that $x \in a \cap b$. Thus $x \in a \wedge x \in b$. Then it follows, of course, that $x \in a$, as desired.

Now let's prove the other half of the biconditional. Thus assume that $b \subseteq a$ and let us prove that $a \cap b = b$. By a previous theorem, it suffices to prove $a \cap b \subseteq b$ and $b \subseteq a \cap b$. The first of these is easy, and does not even use our assumption. So let's prove the second, that $b \subseteq a \cap b$. That is, we must prove that $\forall x[x \in b \rightarrow x \in (a \cap b)]$. This is proven by general conditional proof. Thus, let x be an arbitrary member of b. We need to prove that $x \in a \cap b$. But by our assumption, $b \subseteq a$, so $x \in a$. Hence $x \in a \cap b$, as desired.

Just as there was a question about the official status of the symbol "$\subseteq$," so too there is a question about the official status of the new symbols "$\cap$" and "$\cup$." And again there are two ways to think of them. You can think of them as being like definite descriptions. For example, you can think of "$a \cap b$" as shorthand for "the set whose members are just the objects that are members of both a and b," and use the Russellian analysis of definite descriptions to define them away using the basic set-theoretic language. Alternatively, you can think of them as new binary function symbols added to the language of set theory. In this case, the definitions really amount to new axioms telling us how the function symbols operate.

Remember

Let b and c be sets.

1. $x \in (b \cap c)$ if and only if $(x \in b) \wedge (x \in c)$
2. $x \in (b \cup c)$ if and only if $(x \in b) \vee (x \in c)$

This shows us that $\cap$ is the set-theoretic counterpart of $\wedge$ while $\cup$ is the counterpart of $\vee$.

Exercises and Problems

Problem 10. Let $a = \{2, 3, 4, 5\}$, $b = \{2, 4, 6, 8\}$, and $c = \{3, 5, 7, 9\}$. Compute the following and express your answer in list notation.

1. $a \cap b$
2. $b \cap a$
3. $a \cup b$
4. $b \cap c$
5. $b \cup c$
6. $(a \cap b) \cup c$
7. $a \cap (b \cup c)$

Problem 11. Prove the results in Theorem 4, except for the one proven in the text. Mention what methods of proof you are using.

Problem 12. Prove that for every set a there is a unique set c such that for all x, $x \in c$ iff $x \notin a$. This set c is called the *absolute complement* of a, and is denoted by $\overline{a}$. [This result will not follow from the axioms we eventually adopt. In fact, it will follow that *no* set has an absolute complement.]

8.5 Sets of sets

The Axiom of Comprehension applies quite generally. In particular, it allows us to form sets of sets. For example, suppose we form the sets $\{0\}$ and $\{0, 1\}$. These sets can themselves be collected together into a set $a = \{\{0\}, \{0, 1\}\}$. More generally, we can prove the following:

Theorem 5. (Unordered Pairs) *For any objects x and y there is a (unique) set $a = \{x, y\}$. In* FOL*:*

$$\forall x \forall y \exists! a \forall w (w \in a \leftrightarrow (w = x \vee w = y))$$

Proof: Let x and y be arbitrary objects. First we prove existence. Let $P(z)$ be the wff $z = x \vee z = y$. Then the set

$$\{z \mid P(z)\}$$

has x and y and nothing more as elements. Uniqueness follows from the Axiom of Extensionality.

It is worth noting that our previous observation about the existence of singletons, which we did not prove then, follows from this result. Thus:

Theorem 6. *For any object x there is a singleton set $\{x\}$.*

Proof: To prove this, apply the previous proposition in the case where $x = y$.

In order for set theory to become as widely used as it has, it was important to find a way to represent order. For example, in high school you learned about the representation of lines and curves as sets of pairs of real numbers. A circle of radius one, centered at the origin, is represented as a set of ordered pairs:

$$\{\langle x, y\rangle \mid x^2 + y^2 = 1\}$$

But sets themselves are unordered. For example, $\{1, 0\} = \{0, 1\}$, by Extensionality. So how are we to represent ordered pairs and other ordered objects?

It turns out that there are many ways to do this. The simplest and most widely used is to represent the ordered pair $\langle x, y\rangle$ by means of the set $\{\{x\}, \{x, y\}\}$. Later, we will ask you to prove that the fundamental property of ordered pairs, namely,

$$\langle x, y\rangle = \langle u, v\rangle \rightarrow (x = u \wedge y = v)$$

holds when we represent them this way. Here we simply point out that the definition makes sense, using the previous two results.

Once we have figured out how to represent ordered pairs, the way is open for us to represent ordered triples, quadruples, etc., as well as to represent relations and functions. We will not go into this here, referring the reader instead to any standard set theory book.

Exercises and Problems

Problem 13. Using the previous two results, let $a = \{2, 3\}$ and let $b = \{a\}$. How many members does a have? How many members does b have? Does $a = b$? That is, is $\{2, 3\} = \{\{2, 3\}\}$?

Problem 14. How many sets are members of the set described below?

$$\{\{\}, \{\{\}, 3, \{\}\}, \{\}\}$$

[Hint: First rewrite this using "$\emptyset$" as a notation for the empty set. Then delete from each description of a set any redundancies.]

Problem 15. Apply the Unordered Pair theorem to $x = y = \emptyset$. What set is obtained? Call this set c. Now apply the theorem to $x = \emptyset, y = c$. Do you obtain the same set or a different set?

Problem 16.**

1. How many elements does the set $\{\{x\}, \{x, y\}\}$ contain if $x \neq y$? How many if $x = y$?
2. Define $\langle x, y\rangle = \{\{x\}, \{x, y\}\}$. How is this definition justified? That is, how do we know that for any x and y there is a unique set $\langle x, y\rangle$?
3. Prove that the fundamental property of ordered pairs holds with this definition:

$$\langle x, y\rangle = \langle u, v\rangle \rightarrow (x = u \wedge y = v)$$

Problem* 17. Building on Problem 16, prove that for any two sets a and b, there is a set of all ordered pairs $\langle x, y\rangle$ such that $x \in a$ and $y \in b$. This set is called the *Cartesian Product* of a and b, and is denoted by $a \times b$.

Problem 18. Suppose that a has three elements and b has five. What can you say about the size of $a \cup b$, $a \cap b$, and $a \times b$? [Hint: in some of these cases, all you can do is give upper and lower bounds on the size of the resulting set.]

8.6 The powerset of a set

Once we get used to the idea that sets can be members of other sets, it is natural to form the set of all subsets of any given set b. The following theorem, which is easy to prove, shows that there is one and only one such set. This set is called the *powerset* of b and denoted $\wp b$ or $\wp(b)$.

Theorem 7. (Powersets) *For any set b there is a (unique) set whose members are just the subsets of b.*

> **Proof:** By the Axiom of Comprehension, we may form the set $c = \{x \mid x \subseteq b\}$. This is the desired set. And by the Axiom of Extensionality, there can be only one such set.

Remember

The powerset of a set b is the set of all its subsets:

$$\wp b = \{a \mid a \subseteq b\}$$

By way of example, let us form the powerset of the set $b = \{2, 3\}$. Thus, we need a set whose members are all the subsets of b. There are four of these. The most obvious two are the singletons $\{2\}$ and $\{3\}$. The other two are the empty set, which is a subset of every set, as we saw in Problem 8, and the set b itself, since every set is a subset of itself. Thus:

$$\wp b = \{\emptyset, \{2\}, \{3\}, \{2,3\}\}$$

Here are some facts about the powerset operation. We will ask you to prove them in the problems.

Theorem 8. *Let a and b be sets.*

1. $b \in \wp b$
2. $\emptyset \in \wp b$
3. $a \subseteq b$ *iff* $\wp a \subseteq \wp b$

It is possible for a set to have some of its own subsets as elements. For example, any set that has the empty set as an element has a subset as an element, since the empty set is a subset of every set. To take another example, the set

$$\{\text{Hoover Tower}\}$$

is both a subset and an element of the set

$$\{\text{Hoover Tower}, \{\text{Hoover Tower}\}\}$$

However, it turns out that no set can have *all* of its subsets as elements.

Theorem 9. *For any set b, it is not the case that $\wp b \subseteq b$.*

Proof: Let b be any set. We want to prove that $\wp b \not\subseteq b$. To prove this, we construct a particular subset of b that is not an element of b. Let

$$c = \{x \mid x \in b \wedge x \notin x\}$$

This set c is clearly a subset of b since it was defined to consist of those members of b satisfying some additional condition. It follows from the definition of the powerset operation, that c is an element of $\wp b$. But we will show that $c \notin b$, by proof by contradiction. Suppose that $c \in b$. Then either $c \in c$ or $c \notin c$. But which? It is not hard to see that neither can be the case. First, suppose that $c \in c$. Then by our definition of c, c is one of those members of b that is left out of c. So $c \notin c$. Next consider the possibility that $c \notin c$. But then c is one of those members of b that satisfies the defining condition for c. Thus $c \in c$. Thus we have proven that $c \in c \leftrightarrow c \notin c$, which is a contradiction. So our assumption that $\wp b \subseteq b$ must be false.

This theorem applies to both finite and infinite sets. The proof shows how to take any set b and find a set c which is a subset of b but not a member of b, namely the set $c = \{x \mid x \in b \text{ and } x \notin x\}$. This is sometimes called the *Russell set for b*, after Bertrand Russell. So what we have proved in the preceding can be restated as:

Theorem 10.. *For any set b, the Russell set for a set b, the set*

$$\{x \mid x \in b \wedge x \notin x\},$$

is a subset of b but not a member of b.

This result is, as we will see, a very important result, one that immediately implies Theorem 9.

Let's compute the Russell set for a few sets. If $b = \{0, 1\}$, then the Russell set for b is just b itself. If $b = \{0, \{0, \{0, \dots\}\}\}$ then the Russell set for b is just $\{0\}$ since $b \in b$. Finally, if $b = \{\text{Hoover Tower}\}$, then the Russell set for b is just b itself.

Exercises and Problems

Problem 19. Compute $\wp\{2, 3, 4\}$. Your answer should have eight distinct elements.

Problem 20. Compute $\wp\{2, 3, 4, 5\}$.

Problem 21. Compute $\wp\{2\}$.

Problem 22. Compute $\wp\emptyset$.

Problem 23. Compute $\wp\wp\{2, 3\}$.

Problem 24. Prove the results stated in Theorem 8.

Problem 25. Here are a number of conjectures you might make. Some are true, but some are false. Prove the true ones, and find examples to show that the others are false.

1. For any set b, $\emptyset \subseteq \wp b$.
2. For any set b, $b \subseteq \wp b$.
3. For any sets a and b, $\wp(a \cup b) = \wp a \cup \wp b$.
4. For any sets a and b, $\wp(a \cap b) = \wp a \cap \wp b$.

Problem 26. What is the Russell set for each of the following sets?

1. $\{\emptyset\}$
2. A set a satisfying $a = \{a\}$
3. A set $\{1, a\}$ where $a = \{a\}$
4. The set of all sets.

8.7 Russell's Paradox

We are now in a position to show that something is seriously amiss with the theory we have been developing. Namely, we can prove the negation of Theorem 9. In fact, we can prove the following which directly contradicts Theorem 9.

Theorem 11. *There is a set c such that* $\wp c \subseteq c$.

> **Proof:** Using the Axiom of Comprehension, there is a universal set, a set that contains everything. That is, let $c = \{x \mid x = x\}$. But then every subset of c is a member of c, so $\wp c$ is a subset of c.

The set c used in the above proof is called the "universal set" and is usually denoted by "V." It is called that because it contains everything as a member, including itself. What we have in fact shown is that the powerset of the universal set both is and is not a subset of the universal set.

Let us look at our contradiction a bit more closely. Our proof of Theorem 9, applied to the special case of the universal set, gives rise to the set

$$Z = \{x \mid x \in V \wedge x \notin x\}$$

This is just the Russell set for the universal set. But the proof of Theorem 9 shows that Z is a member of Z if and only if Z is not a member of Z. This set Z is called the (absolute) Russell set, and the contradiction we have just established is called Russell's Paradox.

It is impossible to overdramatize the impact Russell's Paradox had on mathematics at the turn of the century. Simple as it is, it shook the subject to its foundations. It is just as if in arithmetic we discovered a proof that 23+27=50 and 23+27$\neq$ 50. Or as if in geometry we could prove that the area of a square both is and is not the square of the side. But here we are in just that position. This shows that there is something wrong with our starting assumptions of the whole theory, the two axioms with which we began. There simply is no domain of sets which satisfies these assumptions. This discovery was regarded as a paradox just because it had earlier seemed to most mathematicians that the intuitive universe of sets did satisfy the axioms.

Russell's Paradox is just the tip of an iceberg of problematic results in Cantor's theory. These paradoxes resulted in a wide-ranging attempt to clarify the notion of a set, so that a consistent conception could be found to use in mathematics. There is no one single conception which has completely won out in this effort, but all do seem to agree on one thing. The problem with Cantor's theory is that it is too uncritical in its acceptance of "large" collections like the collection V used in the last proof. What the result shows is that there is no such set. So our axioms must be wrong. We must not be able to use just any old property in forming a set.

There is one sense in which our account of the above is inaccurate. That is in presenting Cantor's theory in an axiomatic way. Actually, the emphasis on the axiomatic method in modern mathematics was greatly influenced by the discovery of the set-theoretic paradoxes. Cantor himself did not carefully formulate the principles which he took to be fundamental. It is only with hindsight, once the theory got into trouble, that people went back and gave a careful analysis of Cantor's basic assumptions, the analysis we have presented in this chapter.

8.8 Modern set theory

The paradoxes of Cantor's set theory show us that our intuitive notion of set is simply inconsistent. We must go back and rethink the assumptions on which the theory rests. However, in doing this rethinking, we do not want to throw out the baby with the bath water.

Which of our two assumptions got us into trouble, Extensionality or Comprehension? If we examine the Russell Paradox closely, we see that it is actually a straightforward refutation of the Axiom of Comprehension. It shows that there is no set determined by the property of not belonging to itself. That is, the following is, on the one hand, a logical truth, but also the negation of an instance of Comprehension:

$$\neg\exists c \forall x(x \in c \leftrightarrow x \notin x)$$

The Axiom of Extensionality is not needed in the derivation of this fact. So it is the Comprehension Axiom which is the problem.

Thus, we see that no coherent conception of set can countenance the Russell set. But why is there no such set? It is not enough to say that the set leads us to a contradiction. We would like to understand why this is so. Various answers have been proposed to this question.

One view, which goes back to the mathematician John von Neumann, is based on a metaphor of size. The intuition is that some predicates have extensions that are "too large" to be successfully encompassed as a whole and treated as a mathematical object. Any attempt to consider it as a completed totality is inadequate, as it always has more in it than can be in any set.

On this view, the collection of all sets, for example, is not itself a set, because it is "too big." Similarly, on this view, the Russell collection of those sets that are not members of themselves is also not a set at all. It is too big to be a set. How do we know? Well, on the assumption that it is a set, we get a contradiction. In other words, what was a paradox in Cantor's theory turns into a proof (by contradiction) that Russell's collection is not a set.

How can we take this intuition and incorporate it into our theory? That is, how can we modify the Comprehension Axiom so as to allow the instances we want, but also to rule out these "large" collections?

The answer is a bit complicated. First, we modify the axiom so that we can only form subsets of previously given sets. Intuitively, if we are given a set a and a wff $P(x)$ then we may form the subset of a given by:

$$\{x \mid x \in a \wedge P(x)\}$$

The idea here is that if a is not "too large" then neither is any subset of it. Formally, we express this by the axiom

$$\forall a \exists b \forall x[x \in b \leftrightarrow (x \in a \wedge P(x))]$$

In this form, the axiom scheme is called the Axiom of Separation. Actually, as before, we need the universal closure of this wff, so that a and all other free variables in $P(x)$ are universally quantified.

This clearly blocks us from thinking we can form the set of all sets. We cannot use the Axiom of Separation to prove it exists. (In fact, we will later show that we can prove it does not exist.) And indeed, it is easy to show that the resulting theory is consistent. (See Problem 32.) However, this axiom is far too restricted. It blocks some of the legitimate uses we made of the Axiom of Comprehension. For example, it blocks the proof that the union of two sets always exists. Similarly, it blocks the proof that the powerset of any set exists. Try to prove any of these and you will see that the Axiom of Separation does not give you what you need.

Modern set theory can be thought of as obtained from Cantor's inconsistent theory by weakening the Axiom of Comprehension to the Axiom of Separation, and then throwing back all the instances of Comprehension that seem intuitively true, when we restrict attention to collections that are not too large to be sets. That is, we must throw back in those instances that got inadvertently thrown out.

We do not have space in this book to go into the development of set theory further. Rather, we will state the basic axioms, give a few problems, and refer the interested student to any standard book on modern set theory. We mention those by Enderton, Levy, and Vaught as good examples.

The most common form of modern axiomatic set theory is known as Zermelo-Frankel set theory, also known as ZF. Here, just for the record, is a list of the axioms of ZF. In stating their FOL versions, we use the abbreviations $\exists x \in y\ P$ and $\forall x \in y\ P$ for $\exists x(x \in y \wedge P)$ and $\forall x(x \in y \rightarrow P)$.

1. Axiom of Extensionality: As above.
2. Axiom of Separation: As above.
3. Unordered Pair Axiom: For any two objects there is a set that has both as elements.
4. Union Axiom: Given any set a of sets, the union of all the members of a is also a set. That is:

$$\forall a \exists b \forall x[x \in b \leftrightarrow \exists c \in a(x \in c)]$$

5. Powerset Axiom: Every set has a powerset.
6. Axiom of Infinity: There is a set of all natural numbers.
7. Axiom of Replacement: Given any set a and any operation F that defines a unique object for each x in a, there is a set

$$\{F(x) \mid x \in a\}$$

   That is, if $\forall x \in a \exists! y P(x,y)$, then there is a set $b = \{y \mid \exists x \in aP(x,y)\}$
8. Axiom of Regularity: No set has a nonempty intersection with each of its own elements. That is:

$$\forall b[b \neq \emptyset \rightarrow \exists y \in b(y \cap b = \emptyset)]$$

Of these axioms, only the Axiom of Regularity is not a direct, straightforward logical consequence of Cantor's theory. (It is technically a consequence, though, since Cantor's theory is inconsistent. After all, everything is a consequence of inconsistent premises.) It is intended to rule out "irregular" sets like $a = \{\{\{\ldots\}\}\}$ which is a member of itself.

We should examine these in turn to see if they hold on the modified conception of set, where we mean by "set" one that is not too large to be comprehended as a fixed totality. We leave this as a topic for classroom discussion. We close the chapter by noting that all the theorems stated in this chapter—with the exception of Theorem 11, of course—are theorems of modern set theory.

Remember

The Axiom of Separation): Modern set theory must give up the Axiom of Comprehension. It is replaced by the Axiom of Separation:

$$\forall a \exists b \forall x[x \in b \leftrightarrow (x \in a \wedge P(x))]$$

plus some of the intuitively correct consequences of the Axiom of Comprehension.

Exercises and Problems

Exercise 27. Write out the remaining axioms from above in FOL.

Problem 28. Use the Axioms of Separation and Extensionality to prove that if any set exists, then the empty set exists.

Problem 29. Try to derive the existence of the absolute Russell set from the Axiom of Separation. Where does the proof break down?

Exercise 30. Verify our claim that all of Theorems 1–10 are provable using the axioms of ZF. (Some of the proofs are trivial in that the theorems were thrown in as axioms. Others are not trivial.)

Problem 31. (There is no universal set)

1. Verify that our proof of Theorem 9 can be carried out using the axioms of ZF.
2. Use (1) to prove there is no universal set.

Problem 32. Prove that the Axiom of Separation and Extensionality are consistent. That is, find a universe of discourse in which both are clearly true. [Hint: consider the domain whose only element is the empty set.]

Problem* 33. Show that the theorem about the existence of $a \cap b$ can be proven using the Axiom of Separation, but that the theorem about the existence of $a \cup b$ cannot be so proven. [Come up with a domain of sets in which the axioms are true but the theorem in question is false.]

Problem 34. (The Union Axiom and $\cup$) Problem 33 shows us that we cannot prove the existence of $a \cup b$ from the Axiom of Separation. However, the Union Axiom of ZF is stronger than this. It says not just that $a \cup b$ exists, but that the union of any set of sets exists.

1. Show how to prove the existence of $a \cup b$ from the Union Axiom. What other axioms of ZF do you need to use?
2. Apply the Union Axiom to show that there is no set of all singletons. [Hint: Use proof by contradiction and the fact that there is no universal set.]

Problem* 35. Prove in ZF that for any two sets a and b, the Cartesian product $a \times b$ exists. The proof you gave in an earlier exercise will probably not work here, but the result is provable.

Problem 36. While $\wedge$ and $\vee$ have set-theoretic counterparts in $\cap$ and $\cup$, there is no absolute counterpart to $\neg$.

1. Use the axioms of ZF to prove that no set has an absolute complement.
2. In practice, when using set theory, this negative result is not a serious problem. We usually work relative to some domain of discourse, and form relative complements. Justify this by showing, within ZF, that for any sets a and b, there is a set $c = \{x \mid x \in a \wedge x \notin b\}$. This is called the *relative complement of* b with respect to a.

Problem* 37. Assume the Axiom) of Regularity. Show that no set is a member of itself. Conclude that, if we assume Regularity, then for any set b, the Russell set for b is simply b itself.

Problem 38. Let $\Omega = \{\{\{\ldots\}\}\}$. Show that $\Omega = \{\Omega\}$. Conclude that if we assume the Axiom of Regularity, then there is no such set as Ω.

Problem* 39. (Consequences of the Axiom of Regularity)

1. Show that if there is a sequence of sets with the following property, then the Axiom of Regularity is false:

$$\ldots \in a_{n+1} \in a_n \in \ldots \in a_2 \in a_1$$

2. Show that in ZF we can prove that there are no sets $b_1, b_2, \ldots$, $b_n, \ldots$, where $b_n = \{n, b_{n+1}\}$.
3. In computer science, a *stream* is defined to be an ordered pair $\langle x, y \rangle$ whose first element is an "atom" and whose second element is a stream. Show that if we work in ZF and define ordered pairs as usual, then there are no streams.

There are alternatives to the Axiom of Regularity which have been explored in recent years. We mention our own favorite, the axiom AFA due to Peter Aczel, among others. Using AFA you can prove that a great many sets exist with properties that contradict the Axiom of Regularity. The name "AFA" stands for "anti-foundation axiom."

9

Induction

In earlier chapters we introduced most of the important methods of proof used in rigorous reasoning. But there is one, extremely important method omitted in our earlier discussion: proof by mathematical induction.

By and large, the methods we discussed earlier lined up fairly nicely with various connectives and quantifiers. The main exceptions were proof by contradiction, which can be used to prove statements of any form (since $\mathsf{S} \Leftrightarrow \neg\neg\mathsf{S}$), and the method of proof for numerical claims. In terms of syntactic form, mathematical induction is typically used to prove statements of the form:

$$\forall x[\mathsf{P}(x) \rightarrow \mathsf{Q}(x)]$$

You will recognize that this is also the form of statements proved using general conditional proof. The method of proof by induction is really a jazzed-up version of the method of general conditional proof, one that works when the statements involve a predicate $\mathsf{P}(x)$ defined in a special way. Specifically, proof by induction is available when the predicate $\mathsf{P}(x)$ is defined by what is called an "inductive definition." For this reason, we need to discuss proof by induction and inductive definitions side by side. We will see that whenever a predicate $\mathsf{P}(x)$ is defined by means of an inductive definition, then proof by induction provides a much more powerful method of proof than ordinary general conditional proof. Before we can discuss either of these, though, we need to distinguish both from yet a third process that is often called "induction."

In science, we use the term "induction" whenever we draw a general conclusion on the basis of a finite number of observations. For example, every day I observe that the sun comes up, that dropped things fall down, and that people smile more when it is sunny. I come to infer that this is always the case: that the sun always comes up, that dropped things always fall, that people are always happier when it is sunny.

Of course there is no strict logical justification for such inferences.

I may have inferred some general law of nature, or I may have simply observed a bunch of facts without any law that backs them up. Tomorrow, people may be happier if it rains. (Indeed, given the current drought in California, that is now true here.) Induction, in this sense, does not guarantee that the conclusion follows necessarily from the premises.

This is all by way of contrast with mathematical induction, where one *is* able to justify a general conclusion, with infinitely many instances, on the basis of a finite proof. How is this possible? The key lies in the inductive definitions that underwrite this method of proof. Induction, in our sense, is a logically valid method of proof, as certain as any we have studied so far.

Usually discussions of induction start with the natural numbers, and with examples of induction on the natural numbers. We will start with other examples, examples which show that mathematical induction is not necessarily about numbers. The only reason it applies to the natural numbers is that they can be specified by means of an inductive definition.

9.1 Some examples of induction

Inductive definitions involve setting things up in a certain methodical, step-by-step manner. Proofs by induction take advantage of the structure that results from such inductive definitions. We begin with a simple analogy.

Dominos

When they were younger, Claire and Max liked to build long chains of dominos, all around the house. Then they would knock down the first and, if things were set up right, the rest would all fall down. Little did they know that in so doing they were practicing induction. Setting up the dominos is like setting up an inductive definition. Knocking them all down is like proving a theorem by induction.

There are two things required to make all the dominos fall over. They must be close enough together that when any one domino falls, it knocks down the next. And then, of course, you need to knock down the first. In a proof by induction, these two steps correspond to what are called the inductive step (getting from one to the next) and the basis step (getting the whole thing started).

Notice that there is no need to have just one domino behind each domino. You can have two, as long as the one in front will knock both of its successors down. In this way you can build quite elaborate designs, branching out here and there, and, when the time is right, you can knock them all down with a single flick of the finger. The same is true, as we'll see, with induction.

Inductive definitions involving strings of symbols

Inductive definitions are used a great deal in logic. In fact, we have been using them implicitly throughout this book. For example, our definitions of the wffs of FOL were really inductive definitions. So was our definition of the set of terms of first-order arithmetic. Both of these definitions started by specifying the simplest members of the defined collection, and then gave rules that told us how to generate "new" members of the collection from "old" ones. This is how inductive definitions work.

Let's look at another example, just to make things more explicit. Suppose we want to study an ambiguous variant of propositional logic. We will take some primitive symbols, say $A_1, \ldots, A_n$, and call these propositional letters. Next, we will build up "wffs" from these using our old friends $\neg, \wedge, \vee, \rightarrow$, and $\leftrightarrow$. But we are going to let the language be ambiguous, unlike FOL, by leaving out all parentheses. How will we do this? To distinguish these strings from wffs, let us call them *ambig-wffs*. Intuitively, what we want to say is the following:

1. Each propositional letter is an ambigwff@ambig-wff.
2. If p is any ambig-wff, so is the string $\neg p$.
3. If p and q are ambig-wffs, so are $p \wedge q$, $p \vee q$, $p \rightarrow q$, and $p \leftrightarrow q$.
4. Nothing is an ambig-wff unless it is generated by repeated applications of (1), (2), and (3).

In this definition, clause (1) specifies the basic ambig-wffs. It is called the *base clause* of the definition. Clauses (2) and (3) tell us how to form new ambig-wffs from old ones. They are called *inductive* clauses. The final clause just informs us that all ambig-wffs are generated by the earlier clauses, in case we thought that the Empire State Building or George Bush or the number 2 might be an ambig-wff.

Remember

An inductive definition consists of

- a *base clause,* which specifies the basic elements of the defined set,
- one or more *inductive clauses,* which tell us how to generate additional elements, and
- a *final clause*, which tells us that all the elements are either basic or generated by the inductive clauses.

Assuming the clauses of our inductive definition as premises, we can easily derive a theorem like:

$$A_1 \vee A_2 \wedge \neg A_3 \text{ is an ambig-wff.}$$

The proof would go as follows.

> **Proof:** First, A_1, A_2, and A_3 are ambig-wffs by clause (1). $\neg A_3$ is an ambig-wff by clause (2). Then $A_2 \wedge \neg A_3$ is an ambig-wff by clause (3). Another use of clause (3) gives us the desired ambig-wff $A_1 \vee A_2 \wedge \neg A_3$. (Can you give a different derivation of this ambig-wff, one that applies $\wedge$ before $\vee$?)

This proof shows us how the inductive definition of the ambig-wffs is supposed to work, but it is not an inductive proof. Now that we have the definition, let's try to prove something about the ambig-wffs using the method of inductive proof. Indeed, let's prove a few things that will help us identify strings that are *not* ambig-wffs.

Consider the string $\neg \vee \rightarrow$. Obviously, this is not an ambig-wff. But how do we know? Well, clause (4) says it has to be formed by repeated applications of clauses (1)–(3). Examining these clauses, it seems obvious that anything you get from them will have to contain at least one propositional letter. But what kind of proof is that? What method are we applying when we say "examining these clauses, it seems obvious that ..."? What we need is a way to prove the following simple fact:

Proposition 1. *Every ambig-wff contains at least one propositional letter.*

Notice that this claim has the form of a general conditional, where the antecedent involves an inductively defined predicate:

$$\forall p[(p \text{ is an ambig-wff}) \rightarrow Q(p)]$$

Here, Q is the property of containing at least one propositional letter. What the method of induction allows us to do is prove just such a claim. The way we do it is by showing two things. First, we show that all the basic ambig-wffs, those specified by clause (1), have the property Q. We call this the *basis* step of our inductive proof. Second, we show that if some "old" members of ambig-wff have the property Q, then so will the "new" members generated from them by the inductive clauses (2) and (3). We call this the *inductive* step of the proof. This is just like knocking down dominos, albeit in reverse: the second part shows that if a domino falls down, so will the next one; the first part tips over the initial domino.

Here is how the inductive proof goes:

> **Proof:** We will prove this proposition by induction on the ambig-wffs. *Basis:* For our basis case, we need to show that all the propositional letters are strings that contain at least one propositional letter. But they do, since they in fact consist of exactly one such letter. *Induction:* Suppose p and q are ambig-

wffs that each contain at least one propositional letter. We want to show that the new ambig-wffs generated from these by clauses (2) and (3) will also contain at least one propositional letter. This is clearly true, since $\neg p$ contains all the propositional letters contained in p, and $p \wedge q$, $p \vee q$, $p \rightarrow q$, and $p \leftrightarrow q$ contain all the propositional letters contained in p and q. By induction, we can thus conclude that all ambig-wffs contain at least one propositional letter.

What ultimately justifies our conclusion is clause (4) of the inductive definition. Since nothing is an ambig-wff except the basic elements and things that can be generated from them by repeated applications of our two rules, we can be sure that all the ambig-wffs have the property in question.

Remember

An inductive proof consists of

- a *basis step,* which shows that the property holds of the basic elements, and
- an *inductive step,* which shows that *if* the property holds of some elements, then it holds of any elements generated from them by the inductive clauses.

Let's try another example. Suppose we wanted to prove that the string $A_1 \neg \rightarrow A_2$ is not an ambig-wff. Again, this is pretty obvious, but to prove it we need to prove a general fact about the ambig-wffs, one that will allow us to conclude that this particular string does not qualify. The following fact would suffice:

Proposition 2. *No ambig-wff has the symbol $\neg$ occurring immediately before one of the binary connectives: $\wedge, \vee, \rightarrow, \leftrightarrow$.*

Once again, note that the desired result has the form of a general conditional claim, where the antecedent is our inductively defined predicate:

$$\forall p[(p \text{ is an ambig-wff}) \rightarrow Q(p)]$$

This time, Q is the property of not having $\neg$ occurring immediately in front of a binary connective. To prove this, we need a basis step and an inductive step. The basis step must show that $Q(p)$ holds for those expressions p that are ambig-wffs in virtue of clause (1), that is, the propositional letters. The inductive step involves two cases, one corresponding to premise (2), the other to premise (3). For (2) we must show that if a string p has property Q, so does $\neg p$. For (3), we need to prove that if p and q are strings with property Q then so are $p \wedge q$, $p \vee q$, $p \rightarrow q$, and $p \leftrightarrow q$. If we can do this, the proof will be completed by induction, thanks to clause (4). After all, since every ambig-wff has to be obtained by repeated applications of (1), (2), and (3), every ambig-wff will have been shown to have the property in question.

However, there is a problem when we try to carry out the details of this proof. Do you see what it is? Think about trying to do either part of the inductive step, either (2) or (3). For example, in case (2), how do we know that just because p has property Q so does $\neg p$? Well, we don't. For example, $\rightarrow A_1$ has property Q but $\neg\rightarrow A_1$ does not. (Find a similar problem with case (3).)

This is an example of the so-called *Inventor's Paradox*. It is not a real paradox, as in the cases of Russell's Paradox, but it is a bit counterintuitive. It turns out that proofs by induction often get stuck, not because you are trying to prove something false, but because you are not aiming high enough. You need to prove more. In this case, what we have to prove to keep the induction from getting stuck is this stronger claim: no ambig-wff either begins with a binary connective, or ends with a negation sign, or has a negation sign immediately preceding a binary connective. So let Q' be this stronger property. It is clear that $\forall p[Q'(p) \rightarrow Q(p)]$. Thus, what we need to prove by induction is that

$$\forall p[(p \text{ is an ambig-wff}) \rightarrow Q'(p)]$$

This turns out to be easy, and is left as a problem.

Exercises and Problems

Problem 1. In the state of Euphoria, the following two principles hold:

1. If it is sunny on one day, it is sunny the next day.
2. It is sunny today.

Prove that it is going to be sunny from now on.

Problem 2. Raymond Smullyan, a famous logician/magician, gives the following good advice: (1) always speak the truth, and (2) each day, say "I will repeat this sentence tomorrow." Prove that anyone who did these two things would live forever.

Problem 3. Give at least two distinct derivations which show that the following is an ambig-wff: $A_1 \rightarrow A_2 \leftrightarrow \neg A_2$.

Problem 4. Prove by induction that no ambig-wff begins with a binary connective, ends with a negation sign, or has a negation sign immediately preceding a binary connective.

Problem 5. Prove that no ambig-wff ever has two binary connectives next to one another. Conclude that $A_1 \rightarrow \vee A_2$ is not an ambig-wff.

Problem 6. Modify the inductive definition of ambig-wff as follows, to define the set of semi-wffs:

1. Each propositional letter is a semi-wff.
2. If p is any semi-wff, so is the string $\neg p)$.
3. If p and q are semi-wffs, so are $(p \land q), (p \lor q), (p \rightarrow q), (p \leftrightarrow q)$.
4. Nothing is a semi-wff except in virtue of repeated applications of (1), (2), and (3).

Prove by induction that every semi-wff has the following property: the number of right parentheses is equal to the number of left parentheses plus the number of negation signs.

Problem 7. A *palindrome* is a string of letters that reads the same back to front as front to back. Consider the following attempt to define these by induction:

1. Each letter in the alphabet is a pal.
2. If α is a pal, so is the result of putting any letter both in front of and in back of α (e.g., $b\alpha b$).
3. Nothing is a pal except in virtue of (1) and (2).

Prove by induction that every pal is a palindrome. Is the converse true? If so, prove it. If not, fix up the above definition so that it becomes true.

Problem 8. (Existential wffs) In this problem we return to a topic raised in Problem 7.20. In that problem we defined an existential sentence as one whose prenex form contains only existential quantifiers. A more satisfactory definition can be given by means of the following inductive definition. The existential wffs are defined inductively by the following clauses:

1. Every atomic or negated atomic wff is existential.
2. If $\mathsf{P}_1, \ldots, \mathsf{P}_n$ are existential, so are $(\mathsf{P}_1 \lor \ldots \lor \mathsf{P}_n)$ and $(\mathsf{P}_1 \land \ldots \land \mathsf{P}_n)$.
3. If P is an existential wff, so is $\exists \nu P$, for any variable ν.
4. Nothing is an existential wff except in virtue of (1)–(3).

Prove the following facts by induction:

- If P is an existential wff, then it is logically equivalent to a prenex wff with no universal quantifiers.
- If P is an existential sentence that is true in some world, then it will remain true if new objects are added to the world. [You will need to prove something a bit stronger to keep the induction going.]

Is our new definition equivalent to our old one? If not, how could it be modified to make it equivalent?

Problem 9. Give a definition of universal wff, just like that of existential wff in the previous problem, but with universal quantifiers instead of existential. State and prove results analogous to the results you proved there. Then show that every universal wff is logically equivalent to the negation of an existential wff.

Problem* 10. In Problem 7.25, you were asked to show that the property of being immediately next to is not expressible in the language of Tarski's World. What made this hard was that you needed to prove something by induction on wffs, and we didn't have this method of proof at our disposal. So we were really asking you to discover it for yourself. By now, it should be easier. Go back and try again, if you were not able to do it before.

9.2 Inductive definitions in set theory

The way we have been stating inductive definitions seems reasonably rigorous. Still, one might wonder about the status of clauses like (4): "Nothing is an ambig-wff unless it can be generated by repeated applications of (1), (2), and (3)." This is quite different in character from the others, since it mentions not just the objects one is defining, but the other clauses of the definition itself. One might also wonder just what is getting packed into the phrase "repeated applications."

One way to see that there is something different about clause (4) is to note that the other clauses are obviously expressible using first-order formulas. For example, if concat is a symbol for the concatenation function (that is, the function that takes two expressions and places the first immediately to the left of the second), then one could express (2) as

$$\forall p\ [\mathsf{ambig\text{-}wff}(p) \rightarrow \mathsf{ambig\text{-}wff}(\mathsf{concat}(`\neg', p))]$$

In contrast, clause (4) is not the sort of thing that can be expressed in FOL.

However, it turns out that if we work within set theory, then we can express inductive definitions with first-order sentences. Here, for example, is a definition of the set of ambig-wffs that uses set theory. The English version of the definition is as follows: The set S of ambig-wffs is the smallest set satisfying the following clauses:

1. Each propositional letter is in S.
2. If p is in S, then so is $\neg p$.
3. If p and q are in S, then so are $p \wedge q$, $p \vee q$, $p \rightarrow q$, and $p \leftrightarrow q$.

It turns out that this can be transcribed into the language of set theory in a straightforward way.

What we have done here is replace the puzzling clause (4) by one that refers to the smallest set satisfying (1)–(3). How does that help? First of all, what do we mean by *smallest?* We mean smallest in the sense of subset: we want a set that satisfies (1)–(3) and is a subset of any other set satisfying (1)–(3). How do we know that there is such a smallest set? We need to prove a lemma, to show that our definition makes sense.[1]

Lemma 3. *If S is the intersection of a collection $\mathcal{X}$ of sets, each of which satisfies (1)–(3), S will also satisfy (1)–(3).*

We leave the proof of this lemma as Problem 11.

As a result of this lemma, we know that if we take the intersection of *all* sets that satisfy (1)–(3), then we will have a set that satisfies (1)–(3). Further, it must be the smallest such set, since when you take the intersection of a bunch of sets, the result is always a subset of all of the original sets.

When we prove that all ambig-wffs are in some set Q, we are really showing that the set S of ambig-wffs is a subset of Q. One way to do this is to show that Q satisfies (1)–(3). If it does, then $S \subseteq Q$ since S is the smallest set satisfying (1)–(3). Our problem in the proof of Proposition 2 was that our set Q did *not* satisfy either (2) or (3). So we had to replace this with a smaller set Q' that did satisfy (1)–(3). Then we had $S \subseteq Q' \subseteq Q$, and so $S \subseteq Q$ as desired.

Exercises and Problems

Problem 11. Prove Lemma 3.

Problem 12. Give an inductive definition of the set of wffs of propositional logic, similar to the above, but putting in the parentheses. That is, the set of wffs should be defined as the smallest set satisfying various clauses. Be sure to verify that there is such a smallest set.

Problem 13. Based on your answer to Problem 12, prove that every wff has the same number of left parentheses as right parentheses.

9.3 Induction on the natural numbers

Many students come away from the study of induction feeling that it has something special to do with the natural numbers. By now, it should be obvious that this method of proof is more general than that. We can prove things about many different kinds of sets using induction. In fact, whenever a set is defined inductively, we can prove general claims about its members using an inductive proof. Still, the natural numbers are one of the simplest and most useful examples to which induction applies.

[1] A "lemma" is an auxiliary result, one that is of little intrinsic interest, but which is needed for some larger end. Lemmas have the same formal status as a theorem or proposition, but are considered less significant. We number them in sequence with theorems, though, which explains why the following is Lemma 3.

Just how are the natural numbers defined? Intuitively, the definition runs as follows:

1. 0 is a natural number.
2. If n is a natural number, then $n+1$ is a natural number.
3. Nothing is a natural number except in virtue of repeated applications of (1) and (2).

In set theory, this definition gets codified as follows. The set $\mathbf{N}$ of natural numbers is the smallest set satisfying:

1. $0 \in \mathbf{N}$
2. If $n \in \mathbf{N}$, then $n+1 \in \mathbf{N}$

Based on this definition, we can prove statements about natural numbers by induction. Suppose we have some set Q of natural numbers and want to prove that the set contains all natural numbers:

$$\forall x[x \in \mathbf{N} \rightarrow x \in Q]$$

If we prove the following two things:

1. $0 \in Q$
2. If $n \in Q$, then $n+1 \in Q$

then we know that $\mathbf{N} \subseteq Q$, since $\mathbf{N}$ is defined to be the smallest set satisfying these clauses. And this is just another way of stating the universal claim we want to prove.

Let's work an example that illustrates induction on the natural numbers.

Proposition 4. *For every natural number n, the sum of the first n natural numbers is $n(n+1)/2$.*

Proof: We want to prove the statement $\forall n(n \in \mathbf{N} \rightarrow Q(n))$, where $Q(n)$ is the statement: the sum of the first n natural numbers is $n(n+1)/2$. We do this by induction. There is a basis step and an inductive step. The basis step requires that we prove that the sum of the first 0 natural numbers is 0, which it is. (If you don't like this case, you might check to see that $Q(1)$ holds. You might even go so far as to check $Q(2)$, although it's not necessary.)
To prove the inductive step, we assume that we have a natural number k for which $Q(k)$ holds, and show that $Q(k+1)$ holds. I.e, we assume that the sum of the first k natural numbers is $k(k+1)/2$. We must show that the sum of the first $k+1$ natural numbers is $(k+1)(k+2)/2$. How do we conclude this? We simply note that the sum of the first $k+1$ natural numbers is $k+1$ greater than the sum of the first k natural numbers. We

already know that this latter sum is simply $k(k+1)/2$. Thus the sum that concerns us is

$$\frac{k(k+1)}{2} + (k+1)$$

Now some simple algebra (namely factoring out the $k+1$ term) shows that this is just

$$(k+1)(\frac{k}{2}+1)$$

which in turn is equal to

$$(k+1)\frac{k+2}{2}$$

the desired result.

Exercises and Problems

Problem* 14. Prove that for all natural numbers $n \geq 2$,

$$(1-\frac{1}{2})(1-\frac{1}{3})\ldots(1-\frac{1}{n}) = \frac{1}{n}$$

Problem* 15. Notice that $1^3 + 2^3 + 3^3 = 36 = 6^2$ and that $1^3 + 2^3 + 3^3 + 4^3 + 5^3 = 225 = 15^2$. Prove that the sum of the first n perfect cubes is a square. [Hint: This is an instance of the inventor's paradox. You will have to prove something stronger than this.]

Proving programs correct

Induction is frequently needed to prove properties of computer programs, especially when they contain iteration, loops, or recursion. Here is a simple example.

The following program is written in the language C. We want to prove that this program has the following property: given an input value n the program prints the values n^2 and $2n+1$. For example, given the input 5, the program will print:

$$n^2 = 25, 2n+1 = 11$$

Here is the program:

```
void Function(int n)
{
int x, y, z;
     x = n;
     z = 1;
     y = 0;

     while (x > 0) {
          y = z + y;
```

```
            z = z + 2;
            x = x - 1;
        }
        printf( "n^2 = %d, 2n+1 = %d", y, z);
}
```

This program might be read as follows:

```
Function( input n varies over integers )
{
x, y, z vary over integers
        Set x = n;
        Set z = 1;
        Set y = 0;

        while (x > 0) {
            let y = z + y;
            let z = z + 2;
            let x = x - 1;
        }
        Print ("n^2 =" y, "2n+1 =" z);
}
```

We want to prove that this program has the input-output behavior we attributed to it. But we cannot prove this by itself. It is another case of the inventor's paradox. To prove what we want, we must prove more. Indeed, to prove our program is correct, we need two lemmas.

Lemma 5. *Given input n, there will be exactly n iterations of the while loop.*

This lemma is proved by induction. The main point is that during each pass through the while loop, x is decremented by 1.

Lemma 6. *After k iterations of the while loop, y and z have the values k^2 and $2k+1$, respectively.*

Once again, the lemma is proved by induction. We leave the actual proofs of these two lemmas as problems.

Exercises and Problems

Problem 16. Prove the two lemmas given above.

Problem* 17. Give other examples in your favorite programming language of programs with loops, and show how you need induction to prove that they have the properties you expect.

Problem 18.** In this chapter we have not presented a real theory of inductive definitions. Rather, we have given several examples of inductive definitions, and shown how they are related to proofs by induction. This is all that most students need. There is, however, an extensive theory of inductive definitions. In this problem we sketch the start of this theory. The reader interested in pursuing it further should consult Aczel's chapter on inductive definitions in *The Handbook of Mathematical Logic*.

Let D be some set. By a *monotone operator on* D we mean a function Γ which assigns to each $X \subseteq D$ a subset $\Gamma(X) \subseteq D$ satisfying the following "monotonicity" condition:

$$\forall X \forall Y [X \subseteq Y \rightarrow \Gamma(X) \subseteq \Gamma(Y)]$$

We can think of any of our inductive definitions as given by such a monotone operator, and we can think of any such monotone operator as giving us an inductive definition. The first four parts of this problem give examples of monotone operators, the remainder explores the relation between such operators and inductive definitions.

1. Let D be the set of real numbers and let $\Gamma(X) =$

   $$X \cup \{x + 1 \mid x = 0 \vee x \in X\}$$

   Show that Γ is a monotone operator. Show that $\Gamma(\mathbf{N}) = \mathbf{N}$, where $\mathbf{N}$ is the set of natural numbers. Show that if $\Gamma(X) = X$ then $\mathbf{N} \subseteq X$.
2. Let D be the set of strings made up from the connectives and the propositional letters used in the definition of the ambig-wffs. Let $\Gamma(X)$ be the set of all propositional letters, together with anything obtained by one application of the connectives to something in X. Thus, if $A_1A_1 \in X$ then the following would all be members of $\Gamma(X)$: $\neg A_1A_1$, $A_1A_1 \wedge A_1A_1$, and so forth. Show that Γ is a monotone operator. Show that if S is the set of ambig-wffs, then $\Gamma(S) = S$.
3. Suppose that f is an n-ary function from some set D into D. Let $\Gamma_f(X) =$

   $$X \cup \{f(d_1, \ldots, d_n) \mid d_1, \ldots, d_n \in X\}$$

   Show that Γ is a monotone operator.
4. Let Γ_1 and Γ_2 be monotone operators. Define a new operator by

   $$\Gamma(X) = \Gamma_1(X) \cup \Gamma_2(X)$$

   Show that Γ is also monotone.
5. Now let Γ be any monotone operator on a set D.
   a. Show that $\Gamma(\Gamma(D)) \subseteq \Gamma(D)$
   b. Call a subset X of D Γ-*closed* if $\Gamma(X) \subseteq X$. By the preceding, we know that there is at least one Γ-closed set. (What is it?) Let I be the intersection of all Γ-closed sets. Prove that I is

Γ-closed. This set I is thus the smallest Γ-closed set. It is said to be the set *inductively defined by* the monotone operator Γ.

c. What is the set inductively defined by the operators in each of 1 and 2?

d. Let I be the set inductively defined by Γ. Prove that $\Gamma(I) = I$.

6. Let $I_0 = \emptyset$, $I_1 = \Gamma(I_0)$, $I_2 = \Gamma(I_1)$, ..., $I_{n+1} = \Gamma(I_n)$, for any natural number n. These sets are called the "finite iterates" of Γ. Show (by induction!) that for each n, $I_n \subseteq I$, where I is the set inductively defined by Γ.
7. In many cases, the set inductively defined by Γ is just the union of all these finite iterates. This is the case with all of the inductive definitions we have given in this book. Why? To answer this, prove that if Γ is "finitely based" (in a sense to be defined in a moment), then the set inductively defined by Γ is the union of its finite iterates. Now for the definition. An operator Γ is said to be *finitely based* provided

$$\forall X \forall x[x \in \Gamma(X) \rightarrow \exists Y \subseteq X(Y \text{ finite } \wedge x \in \Gamma(Y))]$$

8. When a monotone operator is not finitely based, one typically has to keep iterating Γ "into the transfinite" if you want to build up the set from below. Trying to make sense of these transfinite inductions was Cantor's original motivation for developing his theory of sets. We have not developed enough set theory to go into this here. But you might like to try to come up with an example of this phenomenon. You will, of course, need to think of an operator that is not finitely based.

Part IV

Advanced Topics

10

Advanced Topics in Propositional Logic

This chapter contains some more advanced ideas and results from propositional logic, logic without quantifiers. Some are quite important in computer science. Others are more theoretical in nature. The chapter can be read following Chapter 4. In fact, the first three sections concern only $\wedge$, $\vee$, and $\neg$, and so can be read following Chapter 3. In this chapter, when we talk about sentences, we always mean propositional sentences, that is, sentences without quantifiers.

10.1 Horn sentences

Recall that in Chapter 3 we showed how to take any sentence built up without quantifiers and transform it into one in conjunctive normal form, CNF, that is, one which is a conjunction of one or more sentences, each of which is a disjunction of one or more literals. Literals are atomic sentences and their negations. We will call a literal *positive* or *negative* depending on whether it is an atomic sentence or the negation of an atomic sentence, respectively.

A particular kind of CNF sentence turns out to be very important in computer science. These are the so-called "Horn" sentences, named not after their shape, but after the American logician Alfred Horn, who first isolated them and studied some of their properties. A *Horn Sentence* is a sentence in CNF that has the following additional property: every disjunction of literals in the sentence contains *at most one* positive literal.[1]

For example, while the following are all in CNF, none of them are Horn sentences:

[1]Later in the chapter we will find that there is a more intuitive way of writing Horn sentences if we use the connective $\rightarrow$. But for now we restrict attention to sentences involving only $\wedge$, $\vee$, and $\neg$.

¬Home(Claire) ∧ (Home(Max) ∨ Happy(Carl))
(Home(Claire) ∨ Home(Max) ∨ ¬Happy(Claire)) ∧ ¬Happy(Carl)
Home(Claire) ∨ Home(Max) ∨ ¬Home(Carl)

The first sentence fails to be a Horn sentence because the second disjunction contains two positive literals, Home(Max) and Home(Carl). The second fails to be a Horn sentence because of the first disjunction. It contains the two positive literals Home(Claire) and Home(Max). Why does the third fail to be a Horn sentence?

By contrast, the following *are* Horn sentences:

¬Home(Claire) ∧ (¬Home(Max) ∨ Happy(Carl))
Home(Claire) ∧ Home(Max) ∧ ¬Home(Carl)
Home(Claire) ∨ ¬Home(Max) ∨ ¬Home(Carl)
Home(Claire) ∧ Home(Max) ∧ (¬Home(Max) ∨ ¬Home(Max))

Examination of each shows that each conjunct contains at most one positive literal as a disjunct. Verify this for yourself to make sure you understand the definition. (Remember that the definition of CNF allows some degenerate cases, as we stressed in Chapter 3.)

For the time being, let us restrict attention to languages where there are no logical dependencies among the atomic sentences. Later, we will discuss how things need to be modified in the general case.

We have already studied the truth table method for checking to see if a sentence is satisfiable or not. This method is completely mechanical, in that you could program a computer to do it. In general, though, the method is quite "expensive." It consumes a lot of resources. For example, a sentence involving 50 atomic sentences has 2^{50} rows in its truth table, a very big number. For Horn sentences, however, we can in effect look at one row. It is this fact that accounts for the importance of this class of sentences. This quick method for checking the satisfiability of Horn sentences, known as the *satisfaction algorithm for Horn sentences*, is really quite simple. We first describe the method, and then apply it to a couple of examples.

Suppose we have a Horn sentence S built out of atomic sentences A_1, ..., A_n. Start out as though you were going to build a truth table, by listing all the atomic sentences in a row, followed by S. But do not put anything under them. The trick is to work back and forth, using the conjuncts of the sentence to figure out which atomic sentences need to have TRUE written beneath them.

To begin, see which if any of the atomic sentences are themselves conjuncts of S. Write TRUE in the reference column under these atomic sentences. If some of the atomic sentences are now assigned TRUE, then use these to fill in as much as you can of the right hand side of the table. For example, if you have written TRUE under A_5, then you will write

FALSE wherever you find $\neg A_5$. This, in turn, may tell you to fill in some more atomic sentences with TRUE. For example, if $\neg A_1 \lor A_3 \lor \neg A_5$ is a conjunct of S, and each of $\neg A_1$ and $\neg A_5$ have been assigned FALSE, then write TRUE under A_3. Proceed in this way until you run out of things to do.

One of two things will happen. One possibility is that you will reach a point where you are forced to assign FALSE to one of the conjuncts of S, and hence to S itself. In this case, the sentence is not satisfiable. But if this does not happen, then S is satisfiable. For then you can fill in all the remaining columns of atomic sentences with FALSE. This will give you a truth assignment that makes S come out true. (There may be other assignments that make S true; our algorithm just generates one of them.)

Let's give two examples, one with each sort of outcome. First, let's consider the sentence

$$\text{Home(Claire)} \land \neg\text{Home(Max)} \land (\text{Home(Max)} \lor \neg\text{Home(Claire)})$$

To make this fit on the page, let's abbreviate the two atomic sentences Home(Claire) and Home(Max) by C and M, respectively. We begin by writing the following table:

C	M	C	$\land$	$\neg$M	$\land$	(M	$\lor$	$\neg$C)

The first step of the above method tells us to put TRUE under any atomic sentence that is a conjunct of S. In this case, this means we should put a TRUE under C. We then see how much of the right side of the table we can fill in. This gives us the following:

C	M	C	$\land$	$\neg$M	$\land$	(M	$\lor$	$\neg$C)
T								F

Looking at the last conjunct, we see that if the whole is to be true, we must also assign TRUE to M.

C	M	C	$\land$	$\neg$M	$\land$	(M	$\lor$	$\neg$C)
T	T							F

But this means the second conjunct gets assigned FALSE, so the whole sentence comes out FALSE.

C	M	C	$\land$	$\neg$M	$\land$	(M	$\lor$	$\neg$C)
T	T			F				F

Thus the sentence is unsatisfiable.

This time, let's look at the sentence

$$(\neg A \lor \neg B) \land (\neg B \lor C) \land B$$

We won't actually write out the table, but instead will just talk through the method. First, we see that if the sentence is to be satisfied, we must assign TRUE to B, since it is a conjunct. Then, looking at the second conjunct, ¬B ∨ C, we see that assigning TRUE to B forces us to assign TRUE to C. But at this point, we run out of things that we are forced to do. So we can assign A the value FALSE, and we will get an assignment that makes the whole sentence TRUE. (Check that this is the case.)

How do we know that this algorithm is correct? Well, we don't, yet. The examples may have convinced you, but they shouldn't have. We really need a proof. But in order to give a proof that it is correct, we need to back up and give a more rigorous definition of what it means for a sentence to be satisfiable. We will do this in the next section, and then sketch a proof that under the new and more rigorous definition, the algorithm is indeed correct.

Exercises and Problems

Problem$_\Diamond$ 1. A sentence in CNF can be thought of as a list of sentences, each of which is a disjunction of literals. In the case of Horn sentences, each of these disjunctions contains at most one positive literal. Open Horn's Sentences. You will see that this is a list of sentences, each of which is a disjunction of literals, at most one of which is positive. Use the algorithm given above to build a world where all the sentences come out true, and save it as World 10.1.

Problem$_\Diamond$ 2. Open Horn's Other Sentences. You will see that this is a list of sentences, each of which is a disjunctive Horn sentence. Use the algorithm given above to see if you can build a world where all the sentences come out true. If you can, save the world as World 10.2. If you cannot, explain how the algorithm shows this.

Problem 3. (Using the algorithm) Try out the above algorithm on the following sentences. If the sentence is not satisfiable, say so. If it is, give the truth value assignment that results from the application of the algorithm. Assume that A, B, C, and D are atomic sentences.

1. A ∧ (¬A ∨ B) ∧ (¬B ∨ C)
2. A ∧ (¬A ∨ B) ∧ ¬D
3. A ∧ (¬A ∨ B) ∧ ¬B
4. ¬A ∧ (¬A ∨ B) ∧ ¬B
5. (¬A ∨ ¬B ∨ C) ∧ (¬A ∨ B) ∧ A ∧ (¬C ∨ ¬B)

Problem* 4. (Special cases of the algorithm) What does the algorithm tell you about a Horn sentence, each of whose conjuncts contain a positive literal? What does it tell you about a Horn sentence, none of whose conjuncts contain a positive literal?

10.2 Satisfiable sentences and logically true sentences, revisited

In Part I of this book, we kept our discussion of truth tables rather informal. That is, we did not give a precise mathematical definition of truth tables. For some purposes this informality suffices. But if we are really going to prove any theorems about FOL, this notion needs to be modeled in a mathematically precise way. For example, if we are to *prove* that the Horn algorithm just presented is correct, we will need to define a mathematically precise model of the intuitive concepts we have been using, including that of the circumstances that make sentences true.

We can abstract away from the particulars of truth tables, and capture what is essential to the notion, as follows. Let us define a *truth assignment* for a first-order language to be any function h from the set of all atomic sentences of that language into the set {TRUE, FALSE}. That is, for each atomic sentence A of the language, h gives us a truth value, written $h(\mathsf{A})$, either TRUE or FALSE. Intuitively, we can think of each such function h as representing one row of the reference column of a large truth table.

Given a truth assignment h, we can define what it means for h to make an arbitrary sentence of the language true or false. There are many equivalent ways to carry out this task. One natural way is to extend h to a function $\hat{h}$ defined on the set of all sentences and taking values in the set {TRUE, FALSE}. Thus if we think of h as giving us a row of the reference column, then $\hat{h}$ fills in the values of the truth tables for all sentences of the language, that is, the values corresponding to h's row. The definition of $\hat{h}$ is what you would expect, given the truth tables.

1. $\hat{h}(\neg \mathsf{Q}) =$ TRUE if and only if $\hat{h}(\mathsf{Q}) =$ FALSE;
2. $\hat{h}(\mathsf{Q} \wedge \mathsf{R}) =$ TRUE if and only if $\hat{h}(\mathsf{Q}) =$ TRUE and $\hat{h}(\mathsf{R}) =$ TRUE;
3. $\hat{h}(\mathsf{Q} \vee \mathsf{R}) =$ TRUE if and only if $\hat{h}(\mathsf{Q}) =$ TRUE or $\hat{h}(\mathsf{R}) =$ TRUE, or both.

With this precise model of a truth assignment, we can give a mathematically precise version of our definitions of satisfiable sentence and logically true sentence. Namely, we say that a first-order sentence S is *satisfiable* provided there is a truth assignment h such that $\hat{h}(\mathsf{S}) =$ TRUE. Similarly, we say that S is *logically true* if every truth assignment h has S coming out true, that is, $\hat{h}(\mathsf{S}) =$ TRUE.[2]

At this stage, we might also point out why so much attention is given

[2]These definitions will need to be amended when we drop the assumption that there are no logical relationships among the atomic sentences of the language, of course, since some of these truth assignments will then represent spurious rows of the truth table.

to showing that a sentence or set of sentences is *not* satisfiable. As we said earlier, one of the central problems in logic is giving an analysis of the notion of logical consequence. Our definition of truth assignments gives us a way to model this notion mathematically for the sentences of propositional logic. Namely, we say that a sentence S is a logical consequence of a set $\mathcal{T}$ of sentences provided every truth assignment that makes all the sentences in $\mathcal{T}$ true also makes S true. In the following, $\mathcal{T} \cup \{\neg S\}$ is the set that has as members all the sentences in $\mathcal{T}$ plus the sentence $\neg S$.

Proposition 1. The sentence S is a logical consequence of the set $\mathcal{T}$ if and only if the set $\mathcal{T} \cup \{\neg S\}$ is not satisfiable.

The proof of this result is left as an exercise.

If $\mathcal{T}$ is finite, we can reduce the question of whether S is a logical consequence of $\mathcal{T}$ to the question of whether a single sentence is not satisfiable, namely the conjunction of the sentences in $\mathcal{T}$ and $\neg S$.

Having given a mathematical model of satisfiability and explained its importance for understanding the notion of logical consequence, we return to prove the result promised in the previous section.

Theorem 2. *The algorithm for the satisfiability of Horn sentences is correct, in that it classifies as satisfiable exactly the satisfiable Horn sentences.*

Proof: There are two things to be proved here. One is that any satisfiable sentence is classified as satisfiable by the algorithm. This half is pretty obvious. For we will only be forced to assign the sentence FALSE in cases in which no possible truth assignment satisfies the sentence. So with satisfiable sentences, we will not be so forced.

The other half of our result is more subtle. Here is the basic idea. Let us assume we are given a Horn sentence S and that our algorithm does not assign FALSE to S. The sentence S is a conjunction of disjunctions of literals, each disjunction having at most one positive literal. To see that S is true under some truth assignment $\hat{h}$, we need only make sure that each conjunct is assigned TRUE under $\hat{h}$. And to do this, we need only make sure that at least one literal in each conjunct is assigned TRUE. If the conjunct is a single atomic sentence, then it is assigned TRUE in the first step. If the conjunct has one positive literal but also some negative literals, then one of two things happens. Either each of the negative literals gets assigned FALSE along the way, in which case the one positive literal then gets assigned TRUE. Or, on the other hand, the whole process terminates in assigning FALSE to an atomic sentence whose negation is one of the literals in the disjunct. The final possibility is that we are dealing with a disjunction of all negative literals. But since we are assuming that S is not assigned FALSE, one of these negative literals is not assigned FALSE along the way, so it gets assigned TRUE at the very end, since at that point the atomic sentence from which it is built is assigned FALSE. This completes the proof.

Recall that in this chapter we have been assuming that there are no logical relationships among the atomic sentences of our language. This restriction is crucial to the result we have just established. For example, consider the sentence

$$\mathsf{Small(b)} \land (\neg\mathsf{Small(b)} \lor \mathsf{Cube(b)}) \land (\neg\mathsf{Cube(b)} \lor \mathsf{Tet(b)})$$

If you apply the algorithm to this sentence, it will claim that the sentence is satisfiable. But upon reflection, we see that the assignment it produces is spurious. Indeed, it is obvious that no world can make this sentence true, for any such world would be one in which b is both a small cube and a small tetrahedron, which is impossible.

The problem lies in a mismatch between our mathematical notion of satisfiability and our intuitive notion. According to the mathematical notion, the sentence displayed *is* satisfiable. One simply takes the truth assignment h which assigns TRUE to the three atomic sentences, Small(a), Cube(a), and Tet(a). But of course this truth assignment does not correspond to a genuine possibility, since nothing can be both a cube and a tetrahedron at the same time.

If there are logical relationships among atomic sentences, we need to modify the algorithm for Horn sentences. At the end of the method, when we were not forced to assign FALSE to the sentence, we assigned all the remaining atomic sentences the value FALSE. The last step in the amended procedure must be a check to see whether that assignment is spurious. If it is, then we will have to search for an alternative, non-spurious assignment, and for that, there may be no algorithmic procedure.

Exercises and Problems

Problem$_\diamond$ 5. Create a sentence file containing five Horn sentences, none of which are true in any world, but which the Horn algorithm classifies as satisfiable. The sentences should take advantage of logical relationships among the predicates of Tarski's World as follows:

1. The relationship between LeftOf and RightOf.
2. The relationship between Small and Large.
3. Logical facts about identity.

4. The relationship between BackOf and Between.
5. The relationship between Large and Larger.

Save your sentences as Sentences 10.5.

10.3 Resolution

People are pretty good at figuring out when one sentence is a logical consequence of another, and when it isn't. If it is, we can usually come up with a proof, especially when we have been taught the important methods of proof. And when it isn't, we can usually come up with an assignment of truth values that makes the premises true and the conclusion false. But for computer applications, we need a reliable and efficient algorithm for determining when one sentence is a consequence of another.

The truth table method provides us with a reliable method. The trouble is that it can be very inefficient. In the case of Horn sentences, we have seen a much more efficient method, one that accounts for the importance of Horn sentences in logic programming. In this section we present a method that applies to arbitrary sentences in CNF. It is not in general as efficient as the Horn sentence algorithm, but it is often much more efficient than brute force checking of truth tables. It also has the advantage that it extends to the full first order language with quantifiers. It is known as the resolution method, and lies at the heart of many applications of logic in computer science. This method, like the Horn sentence algorithm, gives the right results *provided* the atomic sentences of our language are logically independent.

The basic notion in resolution is that of a *set of clauses*. A *clause* is just any finite set of literals. Thus, for example,

$$C_1 = \{\neg\mathsf{Small(a)}, \mathsf{Cube(a)}, \mathsf{BackOf(b,a)}\}$$

is a clause. So is

$$C_2 = \{\mathsf{Small(a)}, \mathsf{Cube(b)}\}$$

The special notation $\Box$ is used for the empty clause. A clause C is said to be satisfied by a truth assignment h provided at least one of the literals in C is assigned TRUE by $\hat{h}$.[3] The empty clause $\Box$ clearly is not satisfiable by any assignment, since it does not contain any elements to be made true. If $C \neq \Box$ then h satisfies C if and only if the disjunction of the sentences in C is assigned true by $\hat{h}$.

[3]Notice that we have now given two incompatible definitions of what it means for an assignment to satisfy a set of literals: one where we defined what it means for an assignment to satisfy a set of sentences thought of as a theory, and one where we think of the set as a resolution clause. It would be better if two different words were used. But they aren't, so the reader must trust to context to tell which use is intended.

A nonempty set $\mathcal{S}$ of clauses is said to be satisfied by the truth assignment h provided each clause C in $\mathcal{S}$ is satisfied by h. Again, this is equivalent to saying that the CNF sentence formed by conjoining the disjunctions formed from clauses in $\mathcal{S}$ is satisfied by $\hat{h}$.

The goal of work on resolution is to come up with as efficient an algorithm as possible for determining whether a set of clauses is satisfiable. The basic insight of the theory stems from the observation that in trying to show that a particular set $\mathcal{S}$ is unsatisfiable, it is often easier to show that a larger set $\mathcal{S}'$ derived from it is unsatisfiable. As long as the method of getting $\mathcal{S}'$ from $\mathcal{S}$ insures that the two sets are satisfied by the same assignments, we can work with the larger set $\mathcal{S}'$. Indeed, we might apply the same method over and over until it became transparent that the sets in question are unsatisfiable. The method of doing this is the so-called resolution method.

The general method of resolution proceeds as follows. We start with a sentence S in CNF which we hope to show is not satisfiable. We transform this sentence into a set $\mathcal{S}$ of clauses in the natural way: replace disjunctions of literals by clauses made up of the same literals, and replace conjunctions by sets of clauses. The aim now is to show $\mathcal{S}$ is *not* satisfiable. To do this, we systematically add clauses to the set in such a way that the resulting set is satisfied by the same assignments as the old set. The new clauses we throw in are called "resolvents" of old clauses. If we can finally get a set of clauses which contains $\square$, and so obviously cannot be satisfied, then we know that our original set $\mathcal{S}$ could not be satisfied.

Before giving the official definition of a resolvent, we give a couple examples. Let C_1 and C_2 be the clauses displayed earlier. Notice that in order for an assignment h to satisfy the set $\{C_1, C_2\}$, h will have to assign TRUE to at least one of Cube(a), Cube(b), or BackOf(b,a). So let $C_3 = \{\mathsf{Cube(a)}, \mathsf{Cube(b)}, \mathsf{BackOf(b,a)}\}$ be an additional clause. Then the set of clauses $\{C_1, C_2\}$ and $\{C_1, C_2, C_3\}$ are satisfied by exactly the same assignments. The clause C_3 is a resolvent of the first set of clauses.

For another example, let C_1, C_2, and C_3 be the following three clauses:

$$\begin{aligned} C_1 &= \{\mathsf{Home(Max)}, \mathsf{Home(Claire)}\} \\ C_2 &= \{\neg\mathsf{Home(Claire)}\} \\ C_3 &= \{\neg\mathsf{Home(Max)}\} \end{aligned}$$

Notice that in order for an assignment to satisfy both C_1 and C_2, you will have to satisfy the clause:

$$C_4 = \{\mathsf{Home(Max)}\}$$

Thus we can throw this resolvent C_4 into our set. But when we look at $\{C_1, C_2, C_3, C_4\}$, it is obvious that this new set of clauses cannot be satisfied. C_3 and C_4 are in direct conflict. So the original set is not satisfiable.

With these examples in mind, we now define what it means for one clause, say R, to be a resolvent of two other clauses, say C_1 and C_2.

Definition 1. A clause R is a *resolvent* of clauses C_1 and C_2 if there is an atomic sentence in one of the clauses whose negation is in the other clause, and if R is the set of all the other literals in either clause.

Here are some more examples. Assume A, B, C, and D are atomic. We use $\Box$ as above for the empty clause.

$$\frac{\{A, D\} \quad \{\neg A\}}{\{D\}}$$

$$\frac{\{A, \neg A\} \quad \{A\}}{\{A\}}$$

$$\frac{\{B, C\} \quad \{\neg B, \neg D\}}{\{C, \neg D\}}$$

$$\frac{\{D\} \quad \{\neg D\}}{\Box}$$

The key fact about Resolution is expressed in the following theorem.

Theorem 3. *If $\mathcal{S}$ is an unsatisfiable set of clauses in a language with logically independent atomic sentences, then it is always possible, by successive resolutions, to arrive at $\Box$.*

The proof of this theorem will be outlined in a later problem. Here is an example which illustrates it.

Suppose S is the following sentence in CNF:

$$\neg A \land (B \lor C \lor B) \land (\neg C \lor \neg D) \land (A \lor D) \land (\neg B \lor \neg D)$$

Converting the sentence S to a set $\mathcal{S}$ of clauses gives us the following clauses:

$$\{\neg A\},\ \{B, C\},\ \{\neg C, \neg D\},\ \{A, D\},\ \{\neg B, \neg D\}$$

Our aim is to use Resolution to show that this set of clauses (and hence the original sentence S) is not satisfiable.

$$\frac{\dfrac{\{A, D\} \quad \{\neg A\}}{\{D\}} \qquad \dfrac{\dfrac{\{B, C\} \quad \{\neg C, \neg D\}}{\{B, \neg D\}} \qquad \{\neg B, \neg D\}}{\{\neg D\}}}{\Box}$$

Since we are able to start with clauses in $\mathcal{S}$ and resolve to the empty clause, we know that the original sentence S is not satisfiable. This is what is known as a proof by Resolution.

A proof by resolution shows that a sentence, or set of clauses, is not satisfiable. But it can also be used to show that a sentence C is a consequence of premises $P_1, \ldots, P_n$. This depends on the observation, made earlier, that C is a consequence of premises $P_1, \ldots, P_n$ if and only if the sentence $P_1 \land \ldots \land P_n \land \neg C$ is not satisfiable. Thus if, after converting this sentence into CNF, we can show by resolution that it is not satisfiable, we will know that C is a logical consequence of $P_1, \ldots, P_n$.

Exercises and Problems

Problem$_\diamond$ 6. Open Alan Robinson's Sentences. The sentences in this file are not mutually satisfiable in any world. Indeed, the first six sentences are not mutually satisfiable. Show that the first five sentences are mutually satisfiable by building a world in which they are all true. Save this as World 10.6. Go on to show that each sentence from 7 on can be obtained from earlier sentences by resolution, if we think of the disjunction in clausal form. The last "sentence," □, is clearly not satisfiable, so this shows that the first six are not mutually satisfiable.

Problem 7. Give an ordinary proof, not using resolution, that the first six sentences of Alan Robinson's Sentences are not satisfiable.

Problem 8. Construct a proof by Resolution showing that the following CNF sentence is not satisfiable.

$$(A \lor \neg C \lor B) \land \neg A \land (C \lor B \lor A) \land (A \lor \neg B)$$

Problem 9. Construct a proof by Resolution showing that the following sentence is not satisfiable. Since the sentence is not in CNF, you will first have to convert it to CNF.

$$\neg\neg A \land (\neg A \lor ((\neg B \lor C) \land B)) \land \neg C$$

Problem 10. Resolution can also be used to show that a sentence is logically true. To show that a sentence is logically true, we need only show that its negation is not satisfiable. Use Resolution to show that the following sentence is logically true.

$$A \lor (B \land C) \lor (\neg A \land \neg B) \lor (\neg A \land B \land \neg C)$$

Problem 11. Use resolution to show that:

1. $\neg A$ is a logical consequence of the following premises: $\neg B$, $\neg A \lor C$, $\neg(C \land \neg B)$.
2. $A \lor B$ is a logical consequence of: $C \lor A, \neg C$.
3. $A \lor B \lor (\neg A \land \neg B)$ is logically true.
4. $\neg(\neg A \land \neg B) \lor \neg(A \lor B)$ is logically true.
5. $\neg A \lor (B \land C)$ is a logical consequence of: $\neg A \lor B, C \lor \neg(A \land B)$.

Problem* 12. (Soundness of Resolution) Let $\mathcal{S}$ be a nonempty set of clauses.

1. Let C_1 and C_2 be clauses in $\mathcal{S}$ and let R be a resolvent of C_1 and C_2. Show that $\mathcal{S}$ and $\mathcal{S} \cup \{R\}$ are satisfied by the same assignments.
2. Conclude that if $\mathcal{S}$ is satisfiable, then it is impossible to obtain $\square$ by successive applications of resolution.

Problem 13.** (Completeness of Resolution) In this exercise we outline the proof of Theorem 3.

1. Assume that $\mathcal{S}$ is a set of clauses and that the only literals appearing in clauses are A and ¬A. Show that if $\mathcal{S}$ is not satisfiable, then $\square$ is a resolvent of clauses in $\mathcal{S}$.
2. Next, assume that $\mathcal{S}$ is a set of clauses and that the only literals appearing in clauses are A, B, ¬A, and ¬B. Form two new sets of clauses as follows. First, form sets $\mathcal{S}^{\mathsf{B}}$ and $\mathcal{S}^{\neg\mathsf{B}}$ where the first of these consists of all clauses in $\mathcal{S}$ that do not contain B and the second consists of all clauses in $\mathcal{S}$ that do not contain ¬B. Notice that these sets can overlap, since some clauses in $\mathcal{S}$ might not contain either. Assume that $\mathcal{S}$ is not satisfiable, and that h is any truth assignment. Show that if $h(\mathsf{B})$ = TRUE, then h cannot satisfy $\mathcal{S}^{\mathsf{B}}$. Similarly, show that if $h(\mathsf{B})$ = FALSE, then h fails to satisfy $\mathcal{S}^{\neg\mathsf{B}}$.
3. With the same setup as above, we now form new sets of clauses $\mathcal{S}_{\mathsf{B}}$ and $\mathcal{S}_{\neg\mathsf{B}}$. The first results from $\mathcal{S}^{\mathsf{B}}$ by throwing out ¬B from any clauses that contain it. The second results from $\mathcal{S}^{\neg\mathsf{B}}$ by throwing out B from its clauses. Show that the observation made above about h still holds for these new sets. Note, however, that neither B nor ¬B appears in any clause in either of these sets. Hence it follows that no assignment satisfies $\mathcal{S}_{\mathsf{B}}$ and that no assignment satisfies $\mathcal{S}_{\neg\mathsf{B}}$.
4. Still continuing with the same setup, show that if $\mathcal{S}$ is not satisfiable, then $\square$ can be obtained as a resolvent of each of $\mathcal{S}_{\mathsf{B}}$ and $\mathcal{S}_{\neg\mathsf{B}}$. Here the result you obtained in part 1 comes into play.
5. Use this result to show that if $\mathcal{S}$ is not satisfiable then either $\square$ or {¬B} can be obtained as a resolvent from $\mathcal{S}^{\mathsf{B}}$. Show similarly that either $\square$ or {B} can be obtained as a resolvent from $\mathcal{S}^{\neg\mathsf{B}}$.
6. Use this to show that if $\mathcal{S}$ is not satisfiable, then $\square$ can be obtained as an eventual resolvent of $\mathcal{S}$.
7. Now you have shown that any unsatisfiable set $\mathcal{S}$ of clauses built from just two atomic sentences has $\square$ as an eventual resolvent. Can you see how this method generalizes to the case of three atomic sentences? You will need to use your results for one and two atomic sentences.
8. If you have studied the chapter on induction, complete this proof

to obtain a general proof of Theorem 3. Nothing new is involved except induction.

10.4 The conditional form of Horn sentences

The definition of Horn sentences given earlier looks ad hoc. Why is this particular type of CNF sentence singled out as special? In this section, we revisit Horn sentences in light of the connective $\rightarrow$, and see that they are not quite as ad hoc as they may have seemed.

Consider the following sentence:

$$(\mathsf{Home(Claire)} \land \mathsf{Home(Max)}) \rightarrow \mathsf{Happy(Carl)}$$

If we replace $\rightarrow$ by its definition in terms of $\neg$ and $\lor$, and then use DeMorgan's Law, we obtain the following equivalent form:

$$\neg\mathsf{Home(Claire)} \lor \neg\mathsf{Home(Max)} \lor \mathsf{Happy(Carl)}$$

This is a disjunction of literals, with only one positive literal. Recall that Horn sentences are just conjunctions of sentences of this sort.

Here are some more examples. Assume that A, B, C, and D are atomic sentences. If we replace $\rightarrow$ by its definition, and use DeMorgan's laws, we find that each sentence on the left is logically equivalent to the Horn sentence on the right.

$$\begin{array}{rcl} (\mathsf{A} \rightarrow \mathsf{B}) \land ((\mathsf{B} \land \mathsf{C}) \rightarrow \mathsf{D}) & \Leftrightarrow & (\neg\mathsf{A} \lor \mathsf{B}) \land (\neg\mathsf{B} \lor \neg\mathsf{C} \lor \mathsf{D}) \\ ((\mathsf{B} \land \mathsf{C} \land \mathsf{D}) \rightarrow \mathsf{A}) \land \neg\mathsf{A} & \Leftrightarrow & (\neg\mathsf{B} \lor \neg\mathsf{C} \lor \neg\mathsf{D} \lor \mathsf{A}) \land \neg\mathsf{A} \\ \mathsf{A} \land ((\mathsf{B} \land \mathsf{C}) \rightarrow \mathsf{D}) & \Leftrightarrow & \mathsf{A} \land (\neg\mathsf{B} \lor \neg\mathsf{C} \lor \mathsf{D}) \end{array}$$

The "typical" Horn sentence consists of a conjunction of sentences, each of which is a disjunction of several negative literals and one positive literal, say,

$$\neg\mathsf{A}_1 \lor \ldots \lor \neg\mathsf{A}_n \lor \mathsf{B}$$

This can be rewritten using $\land$ and $\rightarrow$ as:

$$(\mathsf{A}_1 \land \ldots \land \mathsf{A}_n) \rightarrow \mathsf{B}$$

As we said, this is the typical case. But there are the important limiting cases, disjunctions with a positive literal but no negative literals, and disjunctions with some negative literals but no positive literal. By a logical sleight of hand, we can in fact rewrite these in the same form. The sleight of hand is achieved by introducing a couple of rather odd atomic sentences, True and False. The first of these is assumed to be always true. The second is always false. (You can also think of them as shorthand for some other atomic sentence with these properties, say $(\mathsf{a} = \mathsf{a})$ and $\neg(\mathsf{a} = \mathsf{a})$, respectively.) Using these,

$$\neg\mathsf{A}_1 \lor \ldots \lor \neg\mathsf{A}_n$$

can be rewritten as:

$$(\mathsf{A}_1 \wedge \ldots \wedge \mathsf{A}_n) \rightarrow \mathsf{False}$$

Similarly, we can rewrite the lone atomic sentence B as True → B.

Remember

Any Horn sentence of propositional logic is logically equivalent to a conjunction of conditional statements of the following three forms, where the A_i and B stand for ordinary atomic sentences:

1. $(\mathsf{A}_1 \wedge \ldots \wedge \mathsf{A}_n) \rightarrow \mathsf{B}$
2. $(\mathsf{A}_1 \wedge \ldots \wedge \mathsf{A}_n) \rightarrow \mathsf{False}$
3. True → B

The resolution method was developed and refined by Alan Robinson and other computer scientists interested in automated theorem proving. The power of this method depends on our ability to easily transform reasoning tasks into conjunctive normal form. In practice, we often find ourselves reasoning from a series of conditional or biconditional premises. In these cases, the transformation to the necessary CNF is often quite natural, since $\mathsf{P} \rightarrow \mathsf{Q}$ is equivalent to $\neg\mathsf{P} \vee \mathsf{Q}$, and $\mathsf{P} \leftrightarrow \mathsf{Q}$ is equivalent to $(\neg\mathsf{P} \vee \mathsf{Q}) \wedge (\mathsf{P} \vee \neg\mathsf{Q})$.

Alternative notation

There is an alternative notation connected with Horn clauses in Prolog. The clause

$$(\mathsf{A}_1 \wedge \ldots \wedge \mathsf{A}_n) \rightarrow \mathsf{B}$$

is frequently written

$$\mathsf{B} :- \mathsf{A}_1, \ldots, \mathsf{A}_n$$

or

$$\mathsf{B} \leftarrow \mathsf{A}_1, \ldots, \mathsf{A}_n$$

and read "B, if A_1 through A_n."

Exercises and Problems

Problem 14. Rewrite the following Horn sentences in conditional form, using the ← notation just discussed. Here, as usual, A, B, and C are taken to be atomic sentences.

1. $\mathsf{A} \wedge (\neg\mathsf{A} \vee \mathsf{B} \vee \neg\mathsf{C}) \wedge \neg\mathsf{C}$
2. $(\neg\mathsf{A} \vee \neg\mathsf{B} \vee \mathsf{C}) \wedge \neg\mathsf{C}$
3. $(\neg\mathsf{A} \vee \mathsf{B}) \wedge (\mathsf{A} \vee \neg\mathsf{B})$

Problem 15. Suppose you have a Horn sentence which can be put into conditional form in a way that does not contain any conjunct of form 3 (in the last **Remember** box). Show that it is satisfiable. Similarly, show that if it can be put into a conditional form that does not contain a conjunct of form 2, then it is satisfiable.

Problem 16. (Some standard forms of reasoning.) Use resolution to show the following, all of which we have seen are valid principles of reasoning.

1. B is a logical consequence of A → B and A. (Modus Ponens)
2. ¬A is a consequence of A → B and $\neg B$. (Modus Tollens)
3. A → C is a consequence of the premises A → B and B → C. (Hypothetical Syllogism)
4. (A ∧ B) → C is a consequence of B → C. (Strengthening the Antecedent)
5. A → (B ∨ C) is a consequence of A → B. (Weakening the Consequent)
6. C ∨ D is a consequence of the premises A ∨ B, A → C, and B → D. (Constructive Dilemma)
7. A ↔ C is a consequence of A ↔ B and B ↔ C. (Transitivity of ↔)

Problem 17. Use resolution to show the following. (You gave ordinary proofs of these earlier. Which sort of proof do you prefer, and why?)

1. A → (B → A) is a logical truth.
2. (A → (B → C)) ↔ ((A ∧ B) → C) is a logical truth. (Hint: If you try to do this directly, it becomes very complicated. Try instead to show, using resolution, that each side of this biconditional is a logical consequence of the other.)
3. A ↔ ¬B is a logical consequence of A ∨ B ∨ C, B → (A → ¬C), and A ↔ C.
4. C ∧ D is a consequence of A ∨ (B ∧ C), ¬E, (A ∨ B) → (D ∨ E), and ¬A.
5. C → B is a consequence of ¬A → B, C → (D ∨ E), D → ¬C, and A → ¬E.

11

Advanced Topics in FOL

This chapter presents some more advanced topics in first-order logic. It presupposes the material up through Chapter 6. It is intended as a transition into more advanced logic courses of various sorts.

The first few sections of the chapter deal with a mathematical framework in which FOL can be treated rigorously. Two later sections deal with unification and resolution, topics of importance in computer science. The section on resolution presupposes the material on resolution from the preceding chapter. The chapter ends with a section discussing the famous Completeness and Incompleteness Theorems of Kurt Gödel.

11.1 First-order structures

In our discussion of propositional logic, we started out with the intuitive idea of truth tables. Later, to make this idea more rigorous, we abstracted away many of the specifics of truth tables, capturing what was essential by means of the notion of a truth-assignment. Here, we will do something similar for the language with quantifiers.

Set theory and domains of discourse

In our intuitive explanation of the semantics of quantified sentences, we appealed to the notion of a "domain of discourse," and defined truth and satisfaction relative to such a domain. We took this notion to be an intuitive one, familiar both from our experience using Tarski's World, and from our ordinary experience communicating with others. But to prove rigorous results about the notions we have been studying, we need a rigorous account of this basic idea.

First, recall that the notion of a truth assignment used for propositional logic is not adequate once quantifiers have been introduced into the language. Truth assignments give us no mechanism for dealing with the quantifiers, since the truth of the sentences $\forall x P(x)$ and $\exists x P(x)$ is not a simple function of the truth of other sentences. We need to construct a mathematical object that will do two things for us: specify a collection of objects to represent the domain of discourse in question, and tell us how things in this domain are configured (what properties they have, and so forth).

The mathematical objects that represent a particular domain of discourse are usually called "first-order models" or "first-order structures." We will call them structures, since they represent the structure of the domains about which claims are made. These first-order structures play the same role for the full first-order language that is played by the simpler truth-assignments in propositional logic.

Let us begin with a very simple language, a sublanguage of the Tarski's World language. Assume that we have only three predicates, Cube, Larger, and =, and one name, say c. Even with this simple language there are infinitely many sentences. How should we represent, in a rigorous way, the worlds about which we can make claims using this little language?

First, let's look at a very simple example. Consider Mary Ellen's World, which you can find on your Tarski's World disk. This world has three cubes, one of each size, and one small tetrahedron. The cubes are on the third row, arranged from left to right in increasing order of size. The tetrahedron is in the front, right corner. The small cube is named c. Our goal is to construct a mathematical object that represents everything about this world that is relevant to the truth values of sentences in our toy language. Later, we will generalize this to arbitrary first-order languages.

Since sentences are going to be evaluated in Mary Ellen's World, one thing we obviously need to represent is that the world contains four objects. We do this by using a set $D = \{b_1, b_2, b_3, b_4\}$ of four blocks, where b_1 represents the leftmost block, b_2 the next, and so forth. Thus b_4 represents the sole tetrahedron. This set D is said to be the *domain of discourse* of our first-order structure.

To keep first-order structures as clean as possible, we represent only those features of the domain of discourse that are relevant to the truth of sentences in the given first-order language. Given our current sublanguage, there are many features of Mary Ellen's World that are totally irrelevant to the truth of sentences. For example, since we cannot say anything about position, our mathematical structure need not represent any facts about the positions of our blocks. On the other hand, we can say things about size and shape. Namely, we can say that an object is (or is not) a cube and that one object is (or is not) larger than another. So we will need to represent these sorts of facts. We do this by assigning to the predicate Cube a certain subset Cu of the domain of discourse D, namely, the set of cubes. This set is called the *extension* of the predicate Cube in our structure. In modeling the world depicted above, this extension is the set $\{b_1, b_2, b_3\}$. Similarly, to represent facts about the relative sizes of the objects, we assign to the predicate Larger a set La of ordered pairs $\langle x, y \rangle$, where $x, y \in D$. If $\langle x, y \rangle \in La$, then this represents the fact that x is larger than y. So in our example, we would have

$$La = \{\langle b_2, b_1 \rangle, \langle b_3, b_1 \rangle, \langle b_3, b_2 \rangle, \langle b_2, b_4 \rangle, \langle b_3, b_4 \rangle\}$$

La is said to be the *extension* of Larger in the structure.

There is only one more thing we need to make our representation complete, at least as far as the present language is concerned. We have to hook up the individual constant c with the block it names. That is, we need to build into our structure something that tells us that c is a name of block b_1 rather than one of the other blocks. Or, to put it more technically, we need to represent the fact that in our world b_1 is the *referent* of the name c. So we need to pair the name c with the object b_1 that it names. The simplest way to do this in general is to have a function which assigns to each name in the language whatever object in the domain it happens to name. You might call this function the naming function. (The way we actually handle this when we come to the final definition incorporates the naming function in a slightly different way.)

We have neglected to say anything about the identity predicate =. That is because its extension is fixed, once we have the domain D. It is always interpreted as meaning identity, so the extension is just the set of pairs $\langle a, a \rangle$ where $a \in D$. So in this case, it just consists of the set $\{\langle b_1, b_1 \rangle, \langle b_2, b_2 \rangle, \langle b_3, b_3 \rangle, \langle b_4, b_4 \rangle\}$.

Let's stick with our little language a bit longer, but consider how we would want to represent other worlds. In general we need: a domain of discourse D, a subset Cu of D to serve as the extension of the predicate Cube, a set La of pairs from D to serve as the extension of the predicate Larger, and a pairing of the name c with its referent, some element of the domain D of discourse. In order to have one single object to represent the whole world, with all its relevant facts, we package the domain of discourse, the extensions of the predicates, and the referents of the names, all into one mathematical object. Just how this packaging is done is not too important, and different textbooks do it in somewhat different ways. The most elegant packaging, and the one we adopt, is to use a single function $\mathcal{M}$. ("$\mathcal{M}$" stands for "model.") The function $\mathcal{M}$ is defined on the predicates of the language, the names of the language, and the quantifier symbol $\forall$. Such a function is called a *first-order structure* provided the following conditions are satisfied:

1. $\mathcal{M}(\forall)$ is a nonempty set D, called the *domain of discourse* of $\mathcal{M}$.
2. If P is an n-ary predicate symbol of the language then $\mathcal{M}(\mathsf{P})$ is a set of n-tuples $\langle x_1, \ldots, x_n \rangle$ of elements of D. This set is called

the *extension* of P in $\mathcal{M}$. It is required that the extension of the identity symbol consist of all pairs $\langle x, x \rangle$, for $x \in D$.

3. if c is any name of the language, then $\mathcal{M}(\mathsf{c})$ is an element of D, and is called the *referent* of c in $\mathcal{M}$.

Instead of writing $\mathcal{M}(\mathsf{Cube})$, it is more common to write $\mathsf{Cube}^{\mathcal{M}}$, and similarly for the other predicates and names. And it is common to write just $D^{\mathcal{M}}$ for the domain of discourse $\mathcal{M}(\forall)$, or even just D if it is clear from context which structure $\mathcal{M}$ we are talking about.

Exercises and Problems

Exercise 1. Write out a complete description of a first-order structure $\mathcal{M}$ that would represent Mary Ellen's World. This has been done above except for the packaging into a single function.

Problem$_\diamond$ 2. (Simon says) Open Mary Ellen's World. The structure $\mathcal{M}$ that we have used to model this world, with respect to the sublanguage involving only Cube, Larger, and c, is also a good model of many other worlds. What follows is a list of proposed changes to the world. Some of them are allowable changes, in that if you make the change, the model $\mathcal{M}$ still represents the world with respect to this language. Other changes are not. Make the allowable changes, but not the others.

1. Move everything back one row.
2. Interchange the position of the tetrahedron and the large cube.
3. Make the tetrahedron a dodecahedron.
4. Make the large cube a dodecahedron.
5. Make the tetrahedron (or what was the tetrahedron, if you have changed it) large.
6. Add a cube to the world.
7. Add a dodecahedron to the world.

Save your changed world as World 11.2. Now open Mary Ellen's Sentences. Check to see that all these sentences are true in the world you have built. If they are not, you have made some unallowable changes.

Problem 3. In the text we modeled Mary Ellen's World with respect to one sublanguage of Tarski's World. How would our structure have to be modified if we added the following to the language: Tet, Dodec, Between? That is, describe the first-order structure that would represent Mary Ellen's World, in its original state, for this expanded language. [Hint: One of your extensions will be the empty set.]

Problem 4. Consider a first-order language with one binary predicate Outgrabe. Suppose for some reason we are interested in first-order structures $\mathcal{M}$ for this language which have the particular domain $\{Alice, Mad\ Hatter\}$. List all the sets of ordered pairs that could serve as the extension of the symbol Outgrabe. How many would there be if the domain had three elements?

11.2 Spurious structures

In discussing truth assignments, we saw that when there are logical relations among the atomic sentences of our language, then there can be truth-assignments that do not represent genuine possibilities. The same phenomenon appears here: there can be first-order structures that do not represent any real possibility. In such a case, we need to restrict attention to the class of structures that respect the logical relations among the atomic predicates.

Suppose, for example, that we consider a language just like the one we used above, except that we also allow the predicate Tet. Some structures for this language will not correspond to genuine possibilities. If an object b is in both $\mathsf{Tet}^{\mathcal{M}}$ and $\mathsf{Cube}^{\mathcal{M}}$, then the structure $\mathcal{M}$ does not represent any world, since nothing can be both a cube and a tetrahedron. Thus any structure that does not assign disjoint extensions to Cube and Tet will be spurious.

Exercises and Problems

Problem$_\diamond$ 5. (Spurious first-order structures) Let $D = \{b_1, b_2, b_3, b_4, b_5\}$ be a five element set to be used as the domain of discourse in what follows. Let the language under consideration have only the predicates Cube, Tet, and BackOf. For each of the following, determine if the first-order structure described represents some (or many) worlds you can build with Tarski's World. If it does, build one and save it as world World 11.5.x (where x is the number of the item below). If it does not, then explain why it does not.

1. The model $\mathcal{M}$ is specified by
   - $\mathsf{Cube}^{\mathcal{M}} = \{b_1, b_3, b_5\}$
   - $\mathsf{Tet}^{\mathcal{M}} = \{b_2, b_3\}$
   - $\mathsf{BackOf}^{\mathcal{M}} = \emptyset$
2. The model $\mathcal{M}$ is specified by
   - $\mathsf{Cube}^{\mathcal{M}} = \{b_1, b_3, b_5\}$
   - $\mathsf{Tet}^{\mathcal{M}} = \{b_2, b_4\}$
   - $\mathsf{BackOf}^{\mathcal{M}} = \emptyset$
3. The model $\mathcal{M}$ is specified by
   - $\mathsf{Cube}^{\mathcal{M}} = \{b_1, b_3, b_5\}$
   - $\mathsf{Tet}^{\mathcal{M}} = \{b_2, b_4\}$
   - $\mathsf{BackOf}^{\mathcal{M}} = \{\langle b_4, b_1 \rangle\}$
4. The model $\mathcal{M}$ is specified by
   - $\mathsf{Cube}^{\mathcal{M}} = \emptyset$

   - $\mathsf{Tet}^{\mathcal{M}} = \emptyset$
   - $\mathsf{BackOf}^{\mathcal{M}} = \emptyset$

5. The model $\mathcal{M}$ is specified by
   - $\mathsf{Cube}^{\mathcal{M}} = \{b_3, b_5\}$
   - $\mathsf{Tet}^{\mathcal{M}} = \{b_1\}$
   - $\mathsf{BackOf}^{\mathcal{M}} = \{\langle b_1, b_2\rangle, \langle b_2, b_3\rangle, \langle b_3, b_4\rangle, \langle b_4, b_5\rangle\}$

11.3 Truth and satisfaction, revisited

In Chapter 5, we characterized the notion of truth in a domain of discourse rather informally. You will recall that in order to define what it means for a quantified sentence (either $\forall x S(x)$ or $\exists x S(x)$) to be true, we had to have recourse to the notion of satisfaction, what it means for an object b to satisfy a wff S(x) in a domain of discourse. This was defined in terms of what it means for the simpler sentence S(c) to be true, where c was a new name.

Now that we have defined the notion of a first-order structure, we can treat truth and satisfaction more rigorously. The aim here is just to see how our informal treatment looks when you sit down and treat it mathematically. There should be nothing surprising in this section, unless it is that these intuitive notions are a bit complicated to define rigorously.

In our earlier discussion, we explained what it meant for an object b in the domain of discourse to satisfy a wff S(v) with one free variable. That was enough to serve our needs in discussing truth and the game. However, for more advanced work, it is important to understand what it means for some objects to satisfy a wff $P(x_1, \ldots, x_n)$ with n-free variables, for any $n \geq 0$. The case of $n = 0$ is the important special case where there are *no* free variables, that is, where P is a sentence. Let $\mathcal{M}$ be a first-order structure with domain D. A *variable assignment* in $\mathcal{M}$ is, by definition, some (possibly partial) function h defined on a set of variables and taking values in the set D. Thus, for example, if $D = \{a, b, c\}$ then the following would all be variable assignments in $\mathcal{M}$:

1. the function h_1 which assigns b to the variable x.
2. the function h_2 which assigns a, b, and c to the variables x, y, and z, respectively.
3. the function h_3 which assigns b to all the variables of the language.
4. the function h_4 which is the empty function, that is, does not assign values to any variables.

The special case of the empty variable assignment h_4 is important, so we denote it by $h_\perp$.

We say that a variable assignment is *appropriate* for a wff P if all the free variables of P are in the domain of h. Thus the four variable assignments h_1, h_2, h_3, and h_4 listed above would have been appropriate for the following sorts of wffs, respectively:

1. h_1 is appropriate for any wff with free variable x, or with no free variables;
2. h_2 is appropriate for any wff whose free variables are a subset of $\{x, y, z\}$;
3. h_3 is appropriate for any wff at all; and
4. h_4 is appropriate for any wff with no free variables, that is, for sentences, but not for any wffs with free variables.

We next come to the definition of truth by way of satisfaction. The definition we gave earlier required us to define this by means of substituting names for variables. The definition we are about to give ends up being equivalent, but it avoids this detour. It works by defining satisfaction more generally. In particular, we will define what it means for an assignment h to satisfy a wff $P(x_1, \ldots, x_n)$ in $\mathcal{M}$. We will define this by cases, depending on the main connective or quantifier of P, always defining it in terms of subformulas of P. This will reduce the problem gradually to that of atomic wffs, where we say explicitly what it means.

In order to handle the reduction in the case where P starts with a quantifier, we need a way to change a variable assignment. For example, if h is defined on x and we want to say what it means for h to satisfy ∀z Likes(x, z) then we need to be able to take any object b in the domain of discourse and consider the variable assignment which is just like h except that it assigns the value b to the variable z. We will say that h satisfies our wff ∀z Likes(x, z) if and only if every such modified assignment h' satisfies Likes(y, z). To make this a bit easier to say, we introduce the notation "$h[z/b]$" for the modified variable assignment. Thus, in general, $h[v/b]$ is the variable assignment whose domain is that of h plus the variable v and which takes the same values as h, except that on the variable v, the new assignment takes the value b.

Here are a couple examples, harking back to our earlier examples of variable assignments given above:

1. h_1 assigns b to the variable x, so $h_1[y/c]$ assigns b to x and c to y. By contrast, $h_1[x/c]$ assigns a value only to x, the value c.
2. h_2 assigns a, b, c to the variables x, y, and z, respectively. Then $h_2[x/b]$ assigns the values b, b, and c to x, y, and z respectively. The assignment $h_2[u/c]$ assigns the values c, a, b, and c to the variables u, x, y, and z, respectively.
3. h_3 assigns b to all the variables of the language. $h_3[y/b]$ is the same

assignment, h_3, but $h_3[\mathsf{y}/c]$ is different. It assigns c to y and b to every other variable.

4. h_4, the empty function does not assign values to any variables. Then $h_4[\mathsf{x}/b]$ is the function which assigns b to x.

We are now in a position to define what it means for a variable assignment h to satisfy a wff P in the first-order structure $\mathcal{M}$. First, it is always required that h be appropriate for P, that is, be defined for all the free variables of P, and maybe other free variables. Second, there is nothing at all surprising in the following definition. There shouldn't be, anyway, since we are just trying to make precise the intuitive idea of satisfaction of a formula by a sequence of objects. We suggest that the student work through the example at the end of the definition, referring back to the definition as needed, rather than try to read the definition itself right off.

Definition 1. *Definition of satisfaction.* Let P be a wff and let h be an assignment in $\mathcal{M}$ which is appropriate for P.

1. **The atomic case.** Let's take a typical example. Suppose the wff in question is P(x,c,y). Then the variable assignment h must assign elements to the variables x and y, say $h(\mathsf{x}) = a$ and $h(\mathsf{y}) = d$. Likewise, the structure $\mathcal{M}$ assigns the name c a denotation, say b. Relative to this structure and assignment, the wff P(x,c,y) asserts that the triple $\langle a, b, d\rangle$ is in the extension of P, that is, that $\langle a, b, d\rangle \in \mathsf{P}^{\mathcal{M}}$. This is what it means for h to satisfy P(x,c,y) in $\mathcal{M}$. The case of other atomic wffs is handled similarly.
2. **Negation.** Suppose P is $\neg$Q. Then h satisfies P if and only if h does not satisfy Q.
3. **Conjunction.** Suppose P is Q $\land$ R. Then h satisfies P if and only if h satisfies both Q and R.
4. **Disjunction.** Suppose P is Q $\lor$ R. Then h satisfies P if and only if h satisfies Q or R or both.
5. **Universal quantification.** Suppose P is $\forall$vQ. Then h satisfies P if and only if for every $d \in D^{\mathcal{M}}$, $h[\mathsf{v}/d]$ satisfies Q.
6. **Existential quantification.** Suppose P is $\exists$vQ. Then h satisfies P if and only if for some $d \in D^{\mathcal{M}}$, $h[\mathsf{v}/d]$ satisfies Q.

It is customary to write $\mathcal{M} \models \mathsf{P}\ [h]$ to indicate that the variable assignment h satisfies the wff P in the structure $\mathcal{M}$.

Let's work through a very simple example. We take a structure $\mathcal{M}$ with domain $D = \{a, b, c\}$. Let us suppose that our language contains the binary predicate Likes and that the extension of this predicate is the following set of pairs:

$$\mathsf{Likes}^{\mathcal{M}} = \{\langle a, a\rangle, \langle a, b\rangle, \langle c, a\rangle\}$$

That is, a likes itself and b, c likes a, and b likes no one. Let us consider the wff

$$\exists \mathsf{y}(\mathsf{Likes}(\mathsf{x}, \mathsf{y}) \wedge \mathsf{Likes}(\mathsf{y}, \mathsf{y}))$$

with the single free variable x. If the above definition is doing its stuff, it should turn out that an assignment h satisfies this wff just in case h assigns c to the variable x. After all, c is the only individual in the domain who likes someone who likes himself.

Let's examine the definition of satisfaction to see if this is the way it turns out. First, note that h has to assign *some* value to x, since it has to be appropriate for the formula. Let us call this value e; e is one of a, b, or c. Next, we see from the clause for $\exists$ that h satisfies our wff just in case there is some object $d \in D$ such that $h[\mathsf{y}/d]$ satisfies the wff

$$\mathsf{Likes}(\mathsf{x}, \mathsf{y}) \wedge \mathsf{Likes}(\mathsf{y}, \mathsf{y})$$

But $h[\mathsf{y}/d]$ satisfies this wff if and only if it satisfies both conjuncts. Looking at the atomic case, we see that this is true just in case the pairs $\langle e, d\rangle$ and $\langle d, d\rangle$ are in the extension of Likes. But this can only happen if $e = c$ and $d = a$. Thus the only way our original h can satisfy our wff is if it assigns c to the variable x, as we anticipated.

We are now in a position to define what it means for a first-order sentence P (a wff with no free variables) to be true in a structure $\mathcal{M}$.

Definition 2. A sentence P is *true* in a first-order structure $\mathcal{M}$ (a structure for a language in which P can be expressed) if and only if the empty variable assignment $h_\perp$ satisfies P in $\mathcal{M}$. Otherwise P is false in $\mathcal{M}$. We write $\mathcal{M} \models \mathsf{P}$ if P is true in $\mathcal{M}$.

A number of problems are given later to help you understand that this does indeed model the informal, intuitive notion.

Once we have truth, we can define the important notions of logical truth and logical consequence.

Definition 3. Fix a given first-order language, and let $\mathcal{S}$ be the collection of all non-spurious structures for this language. By "structure" below we will mean a first-order structure $\mathcal{M} \in \mathcal{S}$. By "sentence" we will mean a sentence of this language.

1. A sentence P is *logically true* if it is true in every structure.
2. A sentence P is *satisfiable* if it is true in some structure.
3. A sentence P is *falsifiable* if it is false in some structure. Equivalently, if it is not logically true.
4. A sentence Q is a *logical consequence* of a set $T = \{\mathsf{P}_1, \dots\}$ of sentences if every structure that makes all the sentences in T true also makes Q true.

You may have wondered why Tarski's World is so named. It is our way of paying tribute to Alfred Tarski, the logician who played the pivotal role in the development of the semantic conception of logic. It was Tarski who developed the notion of a first-order structure, the notion of satisfaction, and presented the first analysis of truth, logical truth, and logical consequence along the lines we have sketched here.

Remember

1. First-order structures are mathematical models of the domains about which we make claims using FOL.
2. Variable assignments are functions mapping variables into the domain of some first-order structure.
3. A variable assignment satisfies a wff in a structure if, intuitively, the objects assigned to the variables make the wff true in the structure.
4. Using the notion of satisfaction, we can define what it means for a sentence to be true in a structure.
5. Finally, once we have the notion of truth in a structure, we can model the notions of logical truth, and logical consequence.

Exercises and Problems

Problem 6. (Modifying variable assignments.) Suppose $D = \{a, b, c, d\}$ and let h be the variable assignment which is defined only on the variable x and takes value b. Describe explicitly each of the following:

1. $h[\mathsf{y}/c]$
2. $h[\mathsf{x}/c]$
3. $h[\mathsf{z}/b]$
4. $h[\mathsf{x}/b]$
5. $h[\mathsf{x}/c, \mathsf{z}/d]$
6. $(h[\mathsf{x}/c])[\mathsf{x}/d]$

Problem 7. Consider the language with only one binary predicate symbol P and let $\mathcal{M}$ be the structure with domain $D = \{1, 2, 3\}$ and where the extension of P consists of those pairs $\langle n, m \rangle$ such that $m = n+1$. For each of the following wffs, first describe which variable assignments are appropriate for it. Then describe the variable assignments which satisfy it, much the way we described the variable assignments that satisfy the wff $\exists$z Likes(x, z) in the text.

1. P(y,z)
2. $\exists$yP(y, z)
3. $\forall$zP(y, z)

4. P(x,x)
5. ∃x¬P(x, x)
6. ∀xP(x, x)
7. P(x, x) ∨ P(y, z)
8. ∃x(P(x, x) ∨ P(y, z))
9. ∃y(P(x, x) ∨ P(y, z))
10. ∀y∃zP(y, z)
11. ∀y∃yP(y, z)

Now consider the structure $\mathcal{N}$ with the same domain but where the extension of P is the set of those pairs $\langle n, m\rangle$ such that $n \leq m$. How do your answers change?

Problem* 8. Let h be a variable assignment in $\mathcal{M}$ which is appropriate for the wff P. Show that the following three statements are equivalent:

1. h satisfies P in $\mathcal{M}$
2. h' satisfies P in $\mathcal{M}$ for some extension h' of h
3. h' satisfies P in $\mathcal{M}$ for every extension h' of h

Intuitively, this is true because whether a variable assignment satisfies P can depend only on the free variables of P, but it needs a proof. What does this result say in the case where P is a sentence? Express your answer using the concept of truth. [Hint: You will need to prove this by induction on wffs, so this problem presupposes Chapter 9.]

Problem 9. (Truth assignments and first-order structures.) Given a first-order structure $\mathcal{M}$ for a language L, define a truth assignment $g^{\mathcal{M}}$ as follows: for any atomic sentence A, $g^{\mathcal{M}}(\mathsf{A})$ = TRUE if and only if $\mathcal{M} \models \mathsf{A}$. (Here we are using the notation introduced in Definition 2.) Show that the same "if and only if" holds for all sentences of the language that do not contain quantifiers.

Problem$_\diamond$ 10. (An important problem about satisfiability.) Open Skolem's Sentences. You will notice that these sentences are in pairs. Each even numbered sentence is obtained from the preceding sentence by replacing some name with variables and existentially quantifying the variables. The odd numbered sentence logically implies the even numbered sentence which follows it, of course, by existential generalization. The converse does not hold. But something close to it does. To see what, build any world that makes the even numbered sentences all true, but in which the names b, c, d and e are not used. No matter what world you build, you will then be able to name some objects with these names so as to make the odd numbered sentences true. Do this. Try to formulate what is going on in general terms.

11.4 Skolemization

One important role function symbols play in first-order logic is as a way of simplifying (for certain purposes) sentences that have lots of quantifiers nested inside one another. To see an example of this, consider the sentence

$$\forall x \exists y\ \mathsf{Neighbor}(x, y)$$

Given a fixed domain of discourse (represented by a first-order structure $\mathcal{M}$, say) this sentence asserts that every b in the domain of discourse has at least one neighbor c. Let us write this as

$$\mathcal{M} \models \mathsf{Neighbor}(x, y)[b, c]$$

Now if the original quantified sentence is true, then we can pick out, for each b, one of b's neighbors, say his nearest neighbor or favorite neighbor. Let $f(b)$ be this neighbor, so that we have: for every b

$$\mathcal{M} \models \mathsf{Neighbor}(x, y)[b, f(b)]$$

Now what we would like to say is the following, if we had a function symbol f expressing our function f.

$$\mathcal{M} \models \forall x\ \mathsf{Neighbor}(x, \mathsf{f}(x))$$

This would reduce the quantifier string "$\forall x \exists y$" in the original sentence to the simpler "$\forall x$." So we need to expand our first-order language and give ourselves such a function symbol f to use as a name of f.

This important trick is known as *Skolemization*, after the Norwegian logician Thoralf Skolem. The function f is called a *Skolem function* for the original quantified sentence, and the new sentence, the one containing the function symbol but no existential quantifier, is called the *Skolem normal form* of the original sentence.

Remember

(Simplest case of Skolemization) Given a sentence of the form $\forall x \exists y P(x, y)$ in some first-order language, we Skolemize it by choosing a function symbol f not in the language and writing $\forall x P(x, \mathsf{f}(x))$. Every world that makes the Skolemization true also makes the original sentence true. Every world that makes the original sentence true can be turned into one that makes the Skolemization true by interpreting the function symbol f by a function f which picks out, for any object b in the domain, some object c such that they satisfy the wff $P(x, y)$.

Notice that we did not say that a sentence is logically equivalent to its Skolemization. The situation is a little more subtle than that. If our language allowed existential quantification to apply to function symbols, we could get a logically equivalent sentence, namely

$$\exists f \forall x P(x, f(x))$$

This sort of sentence, however, takes us into what is known as second-order logic, which is beyond the scope of this book.

Skolem functions, and Skolem normal form, are very important in advanced parts of logic. We will discuss one application of them later in the chapter, when we sketch how to apply the Resolution Method to FOL with quantifiers.

One of the reasons that natural language does not get bogged down in lots of embedded quantifiers is that there are lots of expressions that act like function symbols, so that we can usually get by with Skolemizations. Possessives, for example, act as very general Skolem functions. We usually think of the possessive "apostrophe s" as indicating ownership, as in *John's car.* But it really functions much more generally as a kind of Skolem function. For example, if we are trying to decide where the group will eat out, then *Max's restaurant* can refer to the restaurant that Max likes best. Or if we we are talking about logic books, we can use *Kleene's book* to refer not to one Kleene owns, but to one he wrote.

Exercises and Problems

Problem 11. Discuss the logical relationship between the following two sentences. [Hint: One is a logical consequence of the other, but they are not logically equivalent.]

$$\forall y \exists z \, \mathsf{ParentOf}(z, y)$$
$$\forall y \, \mathsf{ParentOf}(\mathsf{bestfriend}(y), y)$$

Explain under what conditions the second would be a Skolemization of the first.

Problem 12. Skolemize the following sentence using the function symbol f.

$$\forall z \exists y [(1 + (z \times z)) < y]$$

Which of the following functions on natural numbers could be used as a Skolem function for this sentence:

1. $f(z) = z^2$
2. $f(z) = z^2 + 1$
3. $f(z) = z^2 + 2$
4. $f(z) = z^3$

11.5 Unification of terms

We now turn to a rather different topic, unification, that applies only to languages that contain function symbols. Unification is of crucial importance when we come to extend the resolution method to the full first-order language.

The basic idea behind unification can be illustrated by comparing a couple of claims. Suppose first that Nancy tells you that Max's father drives a Honda, and that no one's grandfather drives a Honda. Now this is not true, but there is nothing logically incompatible about the two claims. Note that if Nancy went on to say that Max was a father (so that Max's father was a grandfather) we could then accuse her of contradicting herself. Contrast Nancy's claim with Mary's, that Max's grandfather drives a Honda, and that no one's father drives a Honda. Mary can be accused of contradicting herself. Why? Because grandfathers are, among other things, fathers.

More abstractly, compare the following pairs, where P is a unary predicate symbol, f and g are unary function symbols, and a is an individual constant.

First pair:	P(f(a)),	∀x¬P(f(g(x)))
Second pair:	P(f(g(a))),	∀x¬P(f(x))

The first pair is a logical possibility. It is perfectly consistent to suppose that the object $f(a)$ has property P but that no object $f(g(b))$ has property P. This can only happen, though, if a is not of the form $g(b)$. By contrast, the second pair is not a logical possibility. Why? Because if ∀x¬P(f(x)) holds, so does the instance where we substitute g(a) for x: ¬P(f(g(a))). But this contradicts P(f(g(a))).

Unification gives us a useful test to see if sets of claims like the above are contradictory or not. You look at the terms involved and see whether they are "unifiable." The terms f(a) and f(g(x)) in the first pair of sentences are not unifiable, whereas the terms in the second pair, f(g(a)) and f(x), are unifiable. What this means is that in the second case there is a way to substitute a term for x so that the results coincide. This agreement produces a clash between the two original sentences. In the first pair of terms, however, there is no such way to make the terms coincide. No way of substituting a term in for the variable x in f(g(x)) is going to give you the term f(a)

These examples motivate the following definition.

Definition 4. We say that two terms t_1 and t_2 are *unifiable* if there is a substitution of terms for some or all of the variables in t_1 and t_2 such that the terms that result from the substitution are syntactically identical terms. Similarly, any set T of terms, of whatever size, is said to be unifiable if there is a single substitution of terms for some or all of the variables that occur in terms of T so that all the resulting terms are identical.

Notice that whether or not terms are unifiable is a purely syntactic notion. It has to do with terms, not with what they denote. The terms father(Max) and father(father(x)) are not unifiable, regardless of whether or not Max is a father. On the other hand, the terms father(father(Max)) and father(y) are unifiable. Just substitute father(Max) in for the variable y. This means that we can decide whether a pair of terms is unifiable without any idea of what the terms happen to stand for.

Let's give a couple more examples of unification. Suppose we have a binary function symbol f and two unary function symbols g and h. Here is an example of three terms that are unifiable. Can you find the substitution that does the trick?

$$f(g(z), x), \quad f(y, x), \quad f(y, h(a))$$

If you said to substitute h(a) for the variable x and g(z) for y you were right. All three terms are transformed into the term f(g(z), h(a)). Are there any other substitutions that would work? Yes, there are. We could plug any term in for z and get another substitution. The one we chose was simplest in that it was the most general. We could get any other from it by means of a substitution.

Here are some examples of pairs, some of which can, others of which cannot, be unified. See if you can tell which are which before reading on.

g(x),	h(y)
h(f(x, x)),	h(y)
f(x, y),	f(y, x)
g(g(x)),	g(h(y))
g(x),	g(h(z))
g(x),	g(h(x))

Half of these go each way. The ones that are unifiable are the second, third, and fifth. The others are not unifiable. The most general unifiers of the three that are unifiable are, in order:

- Substitute f(x, x) for y
- Substitute some other variable z for both x and y
- Substitute h(z) for x

The first pair is not unifiable because no matter what you do, one will always start with g while the other starts with h. Similarly, the fourth pair is not unifiable because the first will always start with a pair of g's, while the second will always start with a g followed by an h. (The reason the last pair cannot be unified is a tad more subtle. Do you see why?)

There is a very general procedure for checking when two (or more) terms are unifiable or not. It is known as the Unification Algorithm. We will not explain it in this book. But once you have done the following problems, you will basically understand how the algorithm works.

Exercises and Problems

Problem 13. Which of the following terms are unifiable with father(x) and which are not? If they are, give the substitution. If they are not, then explain why not.

1. Max
2. father(Claire)
3. mother(Max)
4. father(mother(Claire))
5. father(mother(y))
6. father(mother(x))

Problem 14. Which of the following terms are unifiable with f(x, g(x)) and which are not? If they are, give the most general unifier. If they are not, then explain why not. (Here, as usual, a and b are names, not variables.)

1. f(a, a)
2. f(g(a), g(a))
3. f(g(x), g(g(x)))
4. h(f(a, g(a)))
5. f(f(a, b), g(f(a, b)))

Problem 15. Find a set of four terms which can simultaneously be unified to obtain the term h(f(h(a), g(a))).

Problem* 16. Show that there are an infinite number of different substitutions that unify the following pair of terms. Find one that is most general.

g(f(x, y)), g(f(h(y), g(z)))

Problem 17. How many substitutions are there that unify the following pair?

g(f(x, x)), g(f(h(a), g(b)))

11.6 Resolution, revisited

In this section we discuss in an informal way how Resolution Method for propositional logic can be extended to full first-order logic by combining the tools we have developed above.

The general situation is that you have some first-order premises $P_1, \ldots, P_n$ and a potential conclusion Q. The question is whether Q follows logically from $P_1, \ldots, P_n$. This, as we have seen, is the same as asking whether the sentence

$$P_1 \land \ldots \land P_n \land \neg Q$$

is unsatisfiable. So the general problem can be looked at as determining, of a fixed sentence, say S, of FOL, whether it is unsatisfiable.

The Resolution Method discussed earlier gives a procedure for testing this when the sentence S contains no quantifiers. But interesting sentences do contain quantifiers. Surprisingly, there is a method for reducing the general case to the case where there are no quantifiers.

An overview of this method goes as follows. First, we know that we can always pull all quantifiers in S out in front by logically valid transformations, and so we can assume S is in prenex form.

Call a sentence *universal* if it is in prenex form and all of the quantifiers in it are universal quantifiers. That is, a universal sentence is of the form

$$\forall x_1 \ldots \forall x_n P(x_1, \ldots, x_n)$$

For simplicity, let us suppose that there are just two quantifiers:

$$\forall x \forall y P(x, y)$$

Let's assume that P contains just two names, b and c, and, importantly, that there are no function symbols in P.

We claim that S is unsatisfiable just in case the following quantifier-free sentence S′ is unsatisfiable:

$$P(b, b) \land P(b, c) \land P(c, b) \land P(c, c)$$

Note that we are not saying the two are equivalent. S obviously entails S′, so if S is satisfiable so is S′. S′ does not in general entail S, but it is the case that if S′ is satisfiable, so is S. The reason is fairly obvious. If you have a world that makes S′ true, look at the subworld that just consists of b and c and the relationships they inherit. This little subworld of two objects makes S true.

This neat little observation allows us to reduce the question of the unsatisfiability of the universal sentence S to a sentence of FOL containing no quantifiers, something we know how to solve using the Resolution Method.

There are a couple of caveats though. First, in order for our proof to work, the sentence S must not contain any predicates that have nontrivial logical relations, since that was true of our Resolution Method. For example, it cannot contain =. Second, our proof requires that the sentence not contain any function symbols.

The last point, about no function symbols, is a real pity, since Skolemization gives us a method for taking any prenex sentence S and finding another one that is universal and satisfiable if and only if S is: just replace all the ∃ one by one, left to right, by function symbols. So if we could only generalize the above method to the case where function symbols are allowed, we would have a general method. This is where the Unification Algorithm comes to the rescue. The basic strategy of resolution has to be strengthened a bit.

Resolution method for FOL: Suppose we have sentences S, S', S'', ... and want to show that they are not simultaneously satisfiable. To do this using resolution, we would carry out the following steps.

1. Put each sentence S into prenex form, say
$$\forall x_1 \exists y_1 \forall x_2 \exists y_2 \ldots P(x_1, y_1, x_2, y_2, \ldots)$$
2. Skolemize each of the resulting sentences, say
$$\forall x_1 \forall x_2 \ldots P(x_1, f_1(x_1), x_2, f_2(x_1, x_2), \ldots)$$
3. Put each quantifier free matrix P into conjunctive normal form, say
$$P_1 \wedge P_2 \wedge \ldots \wedge P_n$$
where each P_i is a disjunction of literals.
4. Distribute the universal quantifiers in each sentence across the conjunctions and drop the conjunction signs, ending with a set of sentences of the form
$$\forall x_1 \forall x_2 \ldots P_i$$
5. Change the bound variables in each of the resulting sentences so that no variable appears in two of them.
6. Drop the universal quantifiers and disjunction signs, ending up with a set of resolution clauses.
7. Resolve the set of clauses, using unification to determine clashes between literals.

Rather than explain this in great detail, which would take us beyond the scope of this book, let's look at a few examples.

Example. Suppose you want to show that $\forall x P(x, b)$ and $\forall y \neg P(f(y), b)$ are not jointly satisfiable, that is, that their conjunction is not satisfiable. With this example, we can skip right to step 6, giving us two clauses, each consisting of one literal. Since we can unify x and $f(y)$, we see that these two clauses resolve to $\square$.

Example. Suppose we are told that the following are both true:
$$\forall x(P(x, b) \vee Q(x))$$
$$\forall y(\neg P(f(y), b) \vee Q(y))$$
and we want to derive the sentence,
$$\forall y(Q(y) \vee Q(f(y)))$$

To show that this sentence follows from the first two, we need to show that those sentences together with the negation of this sentence are not simultaneously satisfiable. We begin by putting this negation into prenex form:

$$\exists y(\neg Q(y) \wedge \neg Q(f(y)))$$

We now want to Skolemize this sentence. Since the existential quantifier in our sentence is not preceded by any universal quantifiers, to Skolemize this sentence we replace the variable y by a 0-ary function symbol, that is, an individual constant:

$$\neg Q(c) \wedge \neg Q(f(c)))$$

Dropping the conjunction gives us the following two sentences:

$$\neg Q(c)$$
$$\neg Q(f(c))$$

We now have four sentences to which we can apply step 6. This yields the following four clauses:

1. $\{P(x,b), Q(x)\}$
2. $\{\neg P(f(y),b), Q(y)\}$
3. $\{\neg Q(c)\}$
4. $\{\neg Q(f(c))\}$

Applying resolution to these shows that they are unsatisfiable. Here is a step-by-step derivation of the empty clause.

	Resolvent	*Resolved Clauses*	*Substitution*
5.	$\{Q(y), Q(f(y))\}$	1, 2	f(y) for x
6.	$\{Q(f(c))\}$	3, 5	c for y
7.	$\square$	4, 6	none needed

Example. Let's look at one more example that shows the whole method at work. Consider the two English sentences:

1. *Everyone admires someone who admires them unless they admire Quayle.*
2. *There are people who admire each other, at least one of whom admires Quayle.*

Suppose we want to use Resolution to show that under one plausible reading of these sentences, (2) follows logically from (1). The readings we have in mind are the following, writing A(x,y) for Admires(x,y), and using q for the name Quayle:

$$(S_1) \quad \forall x[\neg A(x,q) \rightarrow \exists y(A(x,y) \wedge A(y,x))]$$
$$(S_2) \quad \exists x \exists y[A(x,q) \wedge A(x,y) \wedge A(y,x)]$$

When you figure out why S_1 logically entails S_2 in Problem 18, you may decide that these are not reasonable translations of the English. But that is beside the point here.

Our goal is to show that S_1 and $\neg S_2$ are jointly unsatisfiable. Now the sentence $\neg S_2$ is equivalent to the following universal sentence, by DeMorgan's Laws:

$$\forall x \forall y(\neg A(x,q) \vee \neg A(x,y) \vee \neg A(y,x))$$

The sentence S_1 is not logically equivalent to a universal sentence, so we Skolemize it. First, note that it is equivalent to the prenex form:

$$\forall x \exists y[A(x,q) \vee (A(x,y) \wedge A(y,x))]$$

Skolemizing, we get the universal sentence

$$\forall x[A(x,q) \vee (A(x,f(x)) \wedge A(f(x),x))]$$

Putting the quantifier-free part of this in conjunctive normal form gives us:

$$\forall x[(A(x,q) \vee A(x,f(x))) \wedge (A(x,q) \vee A(f(x),x))]$$

This in turn is logically equivalent to the conjunction of the following two sentences:

$$\forall x[A(x,q) \vee A(x,f(x))]$$
$$\forall x[A(x,q) \vee A(f(x),x)]$$

Next, we change variables so that no variable is used in two sentences, drop the universal quantifiers, and form clauses from the results. This leaves us with the following three clauses.

1. $\{A(x,q), A(x,f(x))\}$
2. $\{A(y,q), A(f(y),y)\}$
3. $\{\neg A(z,q), \neg A(z,w), \neg A(w,z)\}$

Finally, we apply resolution to derive the empty clause.

	Resolvent	*Resolved Clauses*	Substitution
4.	$\{A(q,f(q))\}$	1, 3	q for w, x, z
5.	$\{A(f(q),q)\}$	2, 3	q for w, y, z
6.	$\{\neg A(q,f(q))\}$	3, 5	f(q) for z, q for w
7.	$\square$	4, 6	none needed

Exercises and Problems

Problem* 18. Give an understandable English proof that S_2 is a logical consequence of S_1.

Problem 19. Show that the sentence given as a prenex form of S_1 really is logically equivalent to it.

Problem 20. There are usually many ways to proceed in Resolution. In our derivation of $\square$ in the last two examples, we chose optimal derivations. Work out different derivations for both.

Problem 21. Use the resolution method to show that the following sentence is a logical truth.

$$\exists x(P(x) \rightarrow \forall y P(y))$$

Exercise$_\diamond$ 22. The sentence shown to be logically true in Problem 21 looks, at first sight, like an implausible candidate for logical truth. To see why claims of this form cannot be falsified, try constructing a world in which the sentence

$$\exists x(\text{Cube}(x) \rightarrow \forall y \text{Cube}(y))$$

is false. Build a world and play the game committed to FALSE. Change the world and try again. What is the strategy being used by the computer?

11.7 Completeness and incompleteness

We close this book with a brief discussion of two very famous and important results about first-order logic. They are the so-called Completeness Theorem and Incompleteness Theorem of FOL. Both are due to the logician Kurt Gödel, no doubt the greatest logician of all time.

Earlier in this book, you learned the main techniques for giving proofs that one statement is a logical consequence of others. There were simple valid reasoning steps and more intricate methods of proof, like proof by contradiction or the method of general conditional proof. But the definition of logical consequence was fundamentally a semantic one: S is a logical consequence of premises $P_1, \ldots, P_n$ if there is no way for the premises to be true without the conclusion also being true. The question arises as to whether the methods of proof we have given are sufficient to prove everything we would like to prove. Can we be sure that if S is a logical consequence of $P_1, \ldots, P_n$, then we can find a proof of S from $P_1, \ldots, P_n$?

The answer is both yes and no. It depends on the first-order language in question. If we have a first-order language in which all structures are legitimate (that is, none are spurious), then there is a precise sense in which the proof methods we have presented will allow us to prove any consequence of any set of premises. This is usually expressed by saying that, for such a language, the methods of proof are *complete.* This is shown by giving a formal system of deduction, like our system $\mathcal{F}$, containing counterparts of these methods of proof. Once a deductive system is described with sufficient precision, one is in a position to prove, about this formal system, that it is complete. This is in fact one of the main benefits of a formal system of deduction over informal methods of proof.

The first such completeness result was given by Gödel in his dissertation in 1929. If you go on to any further course in logic, you will no doubt see such a proof carried out.

Things become much more complicated when one gives up the assumption that there are no spurious models. If there are complicated relations among atomic sentences, or relations between atomic sentences and quantified sentences, or if there is a fixed intended domain of discourse, then the proof of the Completeness Theorem breaks down.

In some cases, it can be patched up. For example, if the relations among atomic sentences can all be summarized by means of fixed sentences, one can always use these as additional premises, and so reduce proofs using the relations to proofs that invoke the additional premises. For example, to capture the fact that the blocks in Tarski's World can have at most one shape, we could invoke additional premises of the form:

$$\forall x[\mathsf{Cube}(x) \rightarrow \neg\mathsf{Tet}(x)]$$

and similarly for the other shape predicates.

In other cases, however, no such move will work. For example, consider the first-order language of arithmetic. If we restrict attention to the intended domain of discourse, then it is known that the methods of proof discussed above are not sufficient to prove all the truths we would like to prove. Not just that, it is known that no effective specification of methods of proof of the kind we have discussed could possibly produce all and only the theorems of arithmetic!

This important, negative result, known as the Gödel Incompleteness Theorem, is a much deeper result than the Completeness Theorem. It tells us that no methods of proof will ever exhaust the semantical notions of logical truth and logical consequence for a rich language like that of first-order arithmetic. This result lies beyond the scope of this book, but we can at least give the reader a hint of how it is established.

Any system of symbols can be represented in a coding scheme like Morse code, where a sequence of dots and dashes, or equivalently, 0's and 1's, is used to represent any given symbol of the system. In this way, any string of symbols of the system can be represented by a string of 0's and 1's. But we can think of any such sequence as representing a number in binary notation. Hence, we can use natural numbers to represent any string of symbols of the system. In this way, statements about sequences of symbols can be represented by analogous statements about natural numbers.

Now suppose we have some fixed proof regime for a first-order language. Proofs will consist of finite strings of symbols. Using the above trick, it turns out that we can express statements about proofs by means of statements about natural numbers. But now suppose our first-order language is itself the language of arithmetic, interpreted in the intended way, as making claims about the natural numbers. We now have the possibility of making statements in this language which represent, by means of our coding scheme, facts about what is provable and what is not.

Gödel showed how to do all this, and went on to show that it is possible in these circumstances to find a particular statement, let us call it G (for Gödel) that claims of itself that it is not provable in the deductive system we started with.

We ask ourselves whether or not the statement G is true. If it were not true, then of course it must be provable, since that is the only way for it to be false. But the proof system is sound. That is, in setting up the proof system, we guarantee that only true statements can ever be proven. So if G is false, it is true, so we would have a statement that is both true and not true, which is impossible. Thus G must be true. But G claims of itself that it is not provable, so that must be true.

In other words, Gödel's Theorem gives us a recipe for starting with any sound proof system for the language of arithmetic and coming up with a true sentence G which is not provable within the system. This result is one of the most important results in logic, one whose consequences are still being explored today. We urge the interested reader to explore it further.

There are many interesting textbooks in logic which could be usefully studied at this point. We mention Enderton's *Mathematical Introduction to Logic*, Melvin Fitting's *First-order Logic and Automated Theorem Proving*, and Boolos and Jeffrey's *Computability and Logic.*

Exercises and Problems

Problem* 23. Gödel's Incompleteness Theorem was inspired by the famous Liar's Paradox, the sentence *This sentence is not true.*

1. Let us assume that this sentence makes an unambiguous claim. Show that the claim is true if and only if it is not true.
2. Conclude that the sentence must not be making an unambiguous claim.
3. One possibility for locating the ambiguity is in a shift in the domain of discourse as the argument proceeds. Discuss this suggestion.

Problem* 24. (Undefinability of Truth) Show that there is no sentence in the language of arithmetic which says of itself that it is not true. Using this, and Gödel's general method, it is easy to see that the following predicate is not definable in the language of arithmetic: $P(n)$ says that n is the code for a true sentence in the language of arithmetic. This is a theorem of Alfred Tarski.

Problem* 25. (Lob's Paradox) Consider the sentence *If this conditional is true, then logic is the most fascinating subject in the world.* Assume that the sentence makes an unambiguous claim.

1. Use the method of conditional proof (and modus ponens) to establish the claim.
2. Use modus ponens to conclude that logic is the most fascinating subject in the world.

Surely a good way to end a logic course.

Appendix A

How to Use Tarski's World

Tarski's World is a Macintosh program designed for use with this book. It lets you represent simple, three-dimensional worlds inhabited by geometric figures of various kinds and sizes, and test first-order sentences to see whether they are true or false in those worlds. In this appendix we describe how to use Tarski's World. A separate appendix, starting on page 296, contains a brief glossary of Macintosh terminology. If you come across some term you do not understand, try looking it up in the glossary.

A.1 The basics

We begin with instructions on how to get started and get stopped, and explain the basic layout of the Tarski's World screen.

Starting Tarski's World

If you are an experienced Macintosh user, find the Tarski's World icon and double-click on it in the standard way. Experienced users can go on to the next section.

If you are new to the Macintosh, you will need to learn how to turn on the machine. We can't help you with that, since different Macintoshes turn on in different ways. Once the machine is on, wait a few moments, until nothing seems to be going on, and insert your Tarski's World disk.

To see the contents of the Tarski's World disk, double-click the disk icon labelled Tarski's World somewhere on the right of the screen.[1] Then find the icon titled "Tarski's World"—it's the one with the universal quantifier (∀) next to a tetrahedron (or pyramid)—and double-click it. After a moment, the Tarski's World application will appear on your screen. At the top of the screen you see the usual Apple, as well as three menu titles: **File, Edit** and **Display**. Pull down each of these menus to see the sorts of things they contain.

[1]If you are not familiar with Macintosh terminology, see the glossary on page 296, for an explanation of such terms as "icon" and "double-click."

The three main windows

There are three windows on the screen. The *World window* is the large area in the upper left titled "Untitled World." It contains a grid on which objects are placed and, on the left, three square buttons showing a tetrahedron, a cube, and a dodecahedron (or soccer ball). Feel free to click on these buttons to see what happens.

The *Sentence window* is the white window across the bottom of the screen. At first it has "Untitled Sentences" in its title bar and the numeral "1" inside. This is where sentences are entered and evaluated to see whether they are true of the world represented in the World window. Click once in the Sentence window to activate it. Feel free to type something in the Sentence window, say, "I'd rather be in Philadelphia."

On the right of the Sentence window you will see the *Evaluation box*, with "Sentence 1" at the top. The title on the evaluation box indicates which sentence of a list of sentences your cursor is in. This is the sentence that will be evaluated when you use the Evaluation box.

Finally, the *Keyboard window* is the window to the right of the World window. This is the window we generally use to enter sentences of first-order logic. Click on the window to activate it. Feel free to play around by clicking on the buttons of this Keyboard.

Opening saved files

Both worlds and sentence lists can be saved as files on your disk. Indeed, many prepackaged world and sentence files come with Tarski's World. To open a saved file, you will use the **Open**... command from the **File** menu.

To open a file, pull down the **File** menu and choose **Open**.... (Alternatively, you can type Command-O, that is, hold down the key with the cloverleaf on it and type O.) If the World window was active when you did this, you will be presented with a list of all the world files and folders that Tarski's World can see. If the Sentence window was active, you will be presented with a list of sentence files and folders. You can get a list of the World files by clicking on the **World files** button at the bottom of the open dialog box. Similarly, you can get the list of sentence files by clicking on the **Sentence files** button.

You will have to navigate around the disk drives and folders to find the file you want to open. This is a standard Macintosh procedure. If you are unfamiliar with the Macintosh, you may need to get some help. All of the world and sentence files that come with Tarski's World are located in the folder called Exercise Files, located on the Tarski's World disk. If you can see this folder's name in the list, double-click on it to open it. You will then be presentented with the list of world or sentence files in this folder.

There are more files than you can see in the displayed list. To see the others, scroll the list by clicking on the downward pointing arrow in the scroll bar to the right of the list. To move the list back up, use the upward pointing arrow. Alternatively, you can drag the little "scroll box" up and down, to see different portions of the list. To open a given file, select the name from the list, and then click **Open**, or simply double-click on the name. Feel free to try this out: open, say, Ackermann's World. If you have made unsaved changes to the world window, you will be asked if you want to save them. Click on **No** for now. We will tell you how to save changes later. Try opening Ackermann's Sentences.

Starting new files

If you want to start a new world or sentence file, simply choose **New...** from the **File** menu (or type Command-N). You will be given a dialog box which lets you say whether you want a new World or new Sentence list. Choose the desired one and click **OK**.

A faster way of doing this is to click on the window's close box: the little box at the far left of the title bar. The window will then revert to the empty and "untitled" state. If you have made any changes to a file, you will be asked if you want to save them.

Saving a file

To save a list of sentences, choose **Save Sentences** (Command-S) or **Save Sentences As...** from the **File** menu. If the sentence list has never been saved before, a dialog box will appear giving you the option of naming the file you are about to create. If you were to hit the return key, or click the **Save** button, the file would be saved with the default name Untitled Sentences. This is not a good idea. You should type in some other name, like Claire's Sentences, and then either hit the return key or click **Save**.

Once a sentence list has been saved, the name of the file appears in the title bar of the Sentence window. If you are working on a named file, the two commands **Save Sentences** and **Save Sentences As...** behave differently. The first will save a new version of the file under the same name, and the old version will be gone. The second gives you a chance to create a new file, with a new name, and keeps the old file, with its name. For this reason, **Save Sentences As...** is the safer of the two options. This is especially useful if you open a prepackaged exercise file, make changes, and then want to save them without writing over the prepackaged file. In that case, you should make whatever changes you want in the file and then give the revised file a new name using **Save Sentences As**. . . .

All of these procedures apply equally to saving worlds. The keyboard equivalent of **Save World** is Command-R.

Saving a file on another disk

You might want to save the files you create on another disk, say for turning in as homework. To do this, you'll need another disk that has been initialized (see the glossary, page 297). Start saving the file in the usual way. But when the dialog box comes up asking you to name the file, first click **Eject**. The Tarski's World disk will be ejected and you can then insert your other disk. Name and save the file as usual.

Quitting Tarski's World

Eventually you will want to leave Tarski's World. To do this, choose **Quit** from the **File** menu. You can also do this by typing Command-Q. If you've made any unsaved changes to the files, Tarski's World will give you a chance to save them. When you get to the Finder, choose **Shut Down** from the **Special** menu.

A.2 Using the World Window

Adding objects

To put an object on the grid, simply click the button on the left that shows the type of object you want. Try this out. Also try holding down the Option key while you click on an object button. This opens up the object's "parameter window," allowing you to specify other characteristics of the object. Click **OK** when you are satisfied with the characteristics.

Naming objects

To name an object already on the grid, double-click on the object (or select it and choose **Edit Selected Object** from the **Edit** menu). This will open the Object Parameter window displaying the object's current shape, size, and names, if any. To give the object a name, simply choose the desired name by clicking in the box next to it. In first-order logic, one object can have several names, but two objects cannot share the same name. Hence Tarski's World lets you give an object more than one name, but once a name is used, that name cannot be assigned to another object. When you have chosen the object's name, click **OK**.

If you know that you want to name an object (or change its size) when you add it to the world, you can save a bit of effort by holding down the Option key while you click the button to add the object. This immediately opens the Object Parameter window so that you can give the object a name or change its size.

Moving objects

To move an object, position the cursor over the object and drag it to the desired position. (That is, move the mouse's arrow over the object, until the arrow turns into an ×. Then, with the button depressed, move the mouse until the object is where you want it.) Note that you cannot place a large block next to another block. If you try, the block will move to the next available space. If you move a block too close to the edge it will fall off.

Sizing and shaping objects

To change an object's size or shape, open the object's parameter window. You do this by double-clicking on the object. The window displays the object's current shape and size. You can alter either or both of these by clicking in the appropriate circles. (If two objects are in immediately adjacent squares, then neither of them can be large. In that case, the **Large** option will be grey, and cannot be selected.) When you've made your choices, click **OK**.

Deleting objects

To delete an object, click on it. It will quiver with anticipation. Then press the backspace key on the (physical) keyboard, or choose **Clear** from the **Edit** menu. You can also drag the object off the edge of the grid and drop it.

Hiding labels

Whenever you name an object, Tarski's World labels the object with its name. Of course, in the real world we only wear name tags at unpleasant social occasions. Like us, objects in Tarski's World can have names without wearing labels. To hide the labels, simply choose **Hide Labels** from the **Display** menu. To redisplay the labels, choose **Show Labels** from the **Display** menu.

2-D view

Labels aren't the only things that can hide. Sometimes a small object can be obscured from view by another object in front of it. To get a bird's eye view of the world, choose **2-D View** from the **Display** menu. To get back to the usual perspective, choose **3-D View** from the **Display** menu. You can also get back and forth between these two views by typing Command-Y.

Objects can be moved, selected, and changed from the 2-D view in exactly the same way as the 3-D view. (You can even change to the 2-D view in the middle of playing the game; sometimes you will have to in order to pick an appropriate object, or to see what Tarski's World is referring to.)

Rotating Worlds

To rotate a world by 90 degrees in either direction, choose **Rotate World Left** (Command-[) or **Rotate World Right** (Command-]) from the **Display** menu. Rotating right means a clockwise rotation, rotating left means a counterclockwise rotation. Such a rotation counts as a change to the world and will be saved when you save the world.

A.3 Using the Keyboard and Sentence Windows

There are two ways to enter formulas into the Sentence window, from the Keyboard window or from the physical keyboard. (From now on we will capitalize the word "Keyboard" whenever referring to the Keyboard window, and use lower case when referring to the Macintosh's physical keyboard.) Most people find it easier to use the Keyboard than the keyboard. So we will begin by describing the use of the Keyboard.

Writing formulas

Tarski's World makes writing first-order formulas quite painless. As you may have noticed while playing with the Keyboard, when you enter a predicate, like Tet or LeftOf, the insertion point locates itself in the appropriate position for entering "arguments"—variables (u, v, w, x, y, z) or individual constants (a, b, c, d, e, f). In such cases, the insertion point is indicated by a thick, vertical line.

What this means is that a sentence like LeftOf(a,b) can be entered into the sentence list with three mouse clicks: first on the LeftOf button, then on the a button, then on the b button. To enter the same thing from the physical keyboard would require 11 keystrokes.

Besides the various symbols used in the language, there are two more buttons in the Keyboard window, a Delete button and a New Sentence button. The latter looks like a curved arrow and is in the lower right corner of the Keyboard. These are not symbols of the language. The first allows you to delete unwanted symbols and spaces from the Sentence window. It works just like the backspace key on the physical keyboard. The second allows you to add a new sentence to your sentence list, right at the insertion point.

In order to allow you to write more readable formulas, Tarski's World treats brackets and braces as completely equivalent to parentheses. Thus, for example, you could write [LeftOf(a, b) ∧ Large(a)] and Tarski's World will read this as (LeftOf(a, b) ∧ Large(a)).

Commenting your sentences

You can add comments to your sentences in a way that will be ignored by the program when it is checking to see if they are well formed or true. You do this by prefacing each line of text you want ignored by either a double quote (") or a semicolon (;). Either of these will cause Tarski's World to ignore all that follows on the same line. To see how to type things like a question mark, or exclamation point, which can be useful in comments, see the table of equivalents used for typing symbols from the keyboard, page 291.

Creating a list of sentences

To create a whole list of sentences, you first enter one sentence, and then choose **Add Sentence After** from the **Edit** menu. You are given a new, numbered line, and can then enter a new sentence.

Instead of choosing **Add Sentence After** from the **Edit** menu, you can do this from the Keyboard window by clicking the New Sentence button (the roundish arrow) or you can do it directly from the keyboard in two ways. You can type Shift-Return (that is, type Return while holding the shift key down) or Command-A.

There is actually a difference between **Add Sentence After** and Shift-Return. This difference arises when the cursor is not at the right end of a sentence. In this case, **Add Sentence After** adds a blank sentence following the sentence in question, whereas Shift-Return breaks the sentence in two at the cursor's position, putting the second half in the new sentence position. This is also how the New Sentence button works.

To insert a new sentence in your list *before* the current sentence, choose **Add Sentence Before** from the **Edit** menu.

Moving from sentence to sentence

You will often need to move from sentence to sentence within a list of sentences. (The reason is that the Evaluation box applies only to one sentence at a time, the one in which the insertion point is present.) You can move the insertion point with the up and down arrow keys (↑, ↓) on the keyboard (unless you have an antique Macintosh) or by clicking on the sentence of interest with the mouse. If you hold down the Option key, the up arrow takes you to the first sentence of the list and the down arrow takes you the last sentence of the list.

The left and right arrow keys (←, →) on the keyboard also move the insertion point, but only within a single sentence. When holding down the Option key, they take you to the beginning and end of the sentence, respectively.

Deleting sentences

To delete a whole sentence and renumber the sentences that remain, choose **Delete Sentence** from the **Edit** menu, or type Command-D. First make sure the insertion point is somewhere in the sentence you want to delete.

Editing sentences

If you are familiar with editing on the Macintosh, you will find that all of the standard editing techniques work in the Sentence window, but only within a single sentence. For those of you not used to the Mac, we will briefly describe these techniques.

To insert symbols in the middle of a formula displayed in the sentence window, move the mouse until the cursor is at the desired position and click. The blinking insertion point will appear, and anything you type from either keyboard will then show up at the insertion point. You can also move the insertion point with the arrow keys.

If you want to delete some part of a sentence, put the insertion point at the beginning or end of the part to be deleted, and drag it until the highlighting covers everything you want to get rid of. This takes a little practice, but you will soon find it very easy. Once the part you want to get rid of is highlighted, you can delete it by choosing Delete from the Keyboard, or by using the backspace key from the keyboard. Alternatively, you can replace the highlighted portion with other symbols by selecting the desired symbols from the Keyboard (or by typing them on the physical keyboard). This will delete the highlighted portion and immediately put the new symbols in its place.

If you double-click in a sentence, it will select the smallest expression containing the insertion point. Triple-clicking will select the entire sentence.

Note that if your highlight includes parts of two sentences, you will not be able to delete or replace the highlighted portion. Tarski's World then thinks you are trying to delete a sentence boundary, and to do this you must instead use the command **Delete Sentence**.

Typing symbols from the keyboard

Sentences can be entered into the Sentence window by typing them on the physical keyboard. When typing the predicates, you don't have to worry about the case (upper or lower) of the letters, though of course you must spell them correctly. You also have to insert your own punctuation: parentheses after the predicate, and commas to separate multiple "arguments" (as in Between(a,x,z)).

To get the logical symbols (or ? or !) use Table 4.

Either the Sentence window or the Keyboard window must be "ac-

TABLE 4
Keyboard equivalents for logical symbols

To GET:	TYPE:
∀	@ or Option-A
∃	? or Option-S
¬	~ or !
∧	^ or &
∨	+ or	
→	$ or >
↔	% or :
=	=
≠	#
?	Option-/
!	Option-1

tive" before typing on the physical keyboard will have any effect. If you type and nothing shows up, that's because the World window is currently the active window. You can recognize the active window by the parallel lines in the title bar. To activate another window, just click in it somewhere. (You can also activate the Keyboard window by choosing **Show Keyboard** from the **Display** menu, or by typing Command-K. This is also, by the way, how you get the Keyboard window back if you happen to close it for some reason.)

Cutting, copying, and pasting

If you want to change the order of the sentences in a list, or copy a sentence from one file to another, use the Macintosh's cut, copy, and paste functions.

If you highlight a string of symbols and then choose **Cut** or **Copy** from the **Edit** menu, the string of symbols is stored on the Macintosh's Clipboard. The difference between the two commands is that **Cut** deletes the highlighted symbols from their present position, while **Copy** leaves them in place. You can't see the contents of the Clipboard, but the symbols will be there until you cut or copy something else to the Clipboard.

Once something is on the Clipboard, it can be pasted anywhere you want it. Just put the insertion point at the desired place and choose **Paste** from the **Edit** menu. A copy of the string of symbols on the Clipboard will be inserted. You can paste several copies at several different points, if you want to.

The **Clear** command on the **Edit** menu simply deletes selected material, whether in the Sentence window or in the World window.

Moving and sizing windows

All of the windows can be moved wherever you want them on the screen. To move them, click in the title bar and drag. To lengthen the Sentence window, you may first have to move it higher on the screen. Lengthening this window can be especially useful if you are writing a very long sentence, and want to see it in its entirety. If you are forced to have a sentence that won't fit in your Sentence window, the scroll bar on the right of the window will allow you to view the obscured portions of the sentence. Simply click on the upward or downward pointing arrows, or drag the scrolling box up or down.

Printing

To print the list of sentences currently displayed in the Sentence window, choose **Print Sentences...** from the **File** menu and then click **OK** in the dialog box that appears. (If your Macintosh is not hooked up to a printer, this probably won't work. If you have trouble getting the logical symbols to print, try turning off background printing.)

To print the world currently displayed in the World window, choose **Print World...** from the **File** menu and then click **OK** in the dialog box that appears. If you would like to enlarge the image of the world, choose **Page Setup...** from the **File** menu. A good setting to use is an enlargement of 180%, printed sideways on the page.

A.4 Using the Evaluation Box

The Evaluation box appears on the right of the Sentence window. It is what ties the Sentence and World windows together.

Checking syntax

As you will learn, only some strings of symbols are grammatically correct, or well-formed, as we say in logic. These expressions are usually called *well-formed formulas*, or *wffs*. And only some of these are appropriate for making genuine claims about the world. These are called *sentences*. Sentences are wffs with no free variables. You will learn about the difference in Chapter 5. In the earlier chapters, you can take "sentence" and "well-formed formula" to mean the same thing.

To see if what you have written in the sentence window is well formed (or a sentence) choose either the **Yes** button or the **No** button next to the question **WFF?** (or next to **Sent?**) in the evaluation box. When you do this, an x will appear where you clicked. You can then choose **Verify**, and Tarski's World will parse your expression, to see if you are right. It marks correct answers with a check, incorrect answers with X.

Verifying truth

Once there is a sentence in the Sentence window and a world in the World window, you can decide whether or not you think the sentence is true, and mark the Evaluation box accordingly. (You must first indicate that the expression is a wff and a sentence before you are allowed to claim that it is either true or false. This is because only sentences can be true or false.) Once you have staked out a claim about the "truth value" of the sentence, choose **Verify**. Tarski's World will evaluate the sentence in the displayed world, and let you know whether your assessment was correct. After you have done this, the truth value of the sentence will be indicated with a T or F to the left of the sentence itself. (If the expression is not grammatical, it will be marked with $*$, the linguist's symbol for an ungrammatical expression.)

Whenever you choose **Verify**, Tarski's World makes a very rough estimate of the maximum time it might take to evaluate your claim. (This estimate is just based on the number of objects in the world and the complexity of the sentence.) When this estimate is more than a few seconds, Tarski's World puts up a dialog box allowing you to cancel the evaluation, if you get tired of waiting. You should be aware that these "worst case" estimates are usually overly pessimistic. Often, the evaluation takes a fraction of the estimated time.

If you want to evaluate the truth of a sentence there is a quicker way than checking all three boxes of the Evaluation box. Just hit the **Enter** key on the numeric keypad at the right of most keyboards. This is equivalent to putting an $\times$ in all three "Yes" boxes and clicking on **Verify**.

Verifying all sentences

You can also verify the truth of an entire list of sentences at one time by choosing **Verify All Sentences** (Command-F) from the **Edit** menu. Tarski's World will check all the sentences in the file and indicate their truth values to the left of the sentence numbers.

A.5 Playing the game

When you stake out a claim about a world with a complex sentence, you are committed not only to the truth of that sentence, but also to claims about its component sentences. For example, if you are committed to the truth of a conjunction $A \wedge B$ (read "A and B") then you are also committed both to the truth of A and to the truth of B. Similarly, if you are committed to the truth of the negation $\neg A$ (read "not A"), then you are committed to the falsity of A.

This simple observation allows us to play a game that reduces complex commitments to more basic commitments. The latter claims are much easier to evaluate. The rules of the game are part of what you will learn in the body of this book. Here, we will explain the kinds of moves you will make in playing the game.

To play the game, you need a guess about the truth value of the current sentence in the current world. This guess is your initial commitment. The game is of most value when this commitment is wrong, even though you won't be able to win in this case.

To start the game, click the **Game** button in the Evaluation box. Tarski's World will begin by reiterating your initial commitment, and you should respond by clicking **OK** or typing Return. At this point, how the game proceeds depends on both the form of the sentence and your current commitment. A summary of the rules can be found in Table 3 on page 122.

Picking objects and sentences

As you see from the game rules, at certain points you will be asked to pick one sentence from a list of sentences. You do this by clicking on the desired sentence and then clicking **OK**, or by double-clicking on the desired sentence.

At other points in the game, you will be asked to pick an object satisfying some formula. You do this by moving the cursor over the desired object and selecting it. Then click **OK**. Tarski's World assigns a name to the chosen object, for example n1, and labels it. (You can also double-click on the object.)

Backing up and giving up

Tarski's World never makes a mistake in playing the game. It will win if it is possible for it to win, that is, if your initial commitment was wrong. However, *you* may make a mistake, and so lose a game you could have won. All it takes is some bad choices along the way. Tarski's World will take advantage of you. It will not tell you that you made a bad move, or that you could have won; it will simply win the game. What this means is that there are two ways for you to lose: if you were wrong in your initial assessment, or if you make a faulty choice in the play of the game. To put this more positively, if you win a game against the computer, then you can be quite sure that your initial assessment of the sentence, as well as all subsequent choices, were correct.

To make up for the edge the computer has, Tarski's World allows you to retract any choices you have made, no matter how far into the game you've gone. So if you think your initial assessment was correct but that you've made a bad choice along the way, you can always retract some moves by clicking on the **Back** button. If your initial assessment really *was* correct, you should, by using this feature, eventually be able to win. If you can't, your initial commitment was probably wrong: check it with **Verify**.

If, halfway through the play of the game, you realize that your assessment was wrong, and understand why, you can avoid some embarrassment by simply quitting the game. Just click the **Quit** button. This ends the game, but does not shut down Tarski's World.

When to play the game

In general, you won't want to play the game with every sentence. The game is most illuminating when you have incorrectly assessed a sentence's truth value, but are not sure why your assessment is wrong. When this happens, you should always play the game without changing your commitment. Tarski's World will win, but in the course of winning, it will usually make clear to you exactly why your assessment was wrong. That's the real value of the game.

You might wonder what happens when you play the game with a correct assessment. In this case, if you play your cards right, you are guaranteed to win. But Tarski's World does not simply give up. At those points in the game when it needs to make choices, it will make them more or less randomly, hoping that you will blunder somewhere along the line. If you do, it will seize the opportunity and win the game. But, as we have noted, you can always renege by backing up.

A.6 Using the Grader

Tarski's World comes with a grading facility, useful for instructors who want to grade a large number of exercise disks. This facility (accessed by choosing **Grade Student Disks...** on the **File** menu) is of little interest to the student user, though the outcome may be of some concern.

This grader is not an answer key, and so is of no use in figuring out the answer to a problem. The Hypercard stack Grader Stack, included on the Tarski's World disk, explains how the grader works and how to build the files and templates needed to grade student exercises. In addition, instructors can write for a special instructor's disk that contains all the files needed to grade the Tarski's World exercises contained in this book. To receive the instructor's disk, write to CSLI Publications on departmental letterhead. Please indicate that you are using the Macintosh version of the program.

Appendix B

Macintosh Terminology

The following is a brief glossary of some Macintosh terminology used in this book.

Active window At any time, only one window on screen is "active." This is indicated by the lines in its title bar.

Application A computer program designed for doing a particular kind of thing. For example, Tarski's World is an application, as is the Finder, from which you start up Tarski's World. You are probably also familiar with word processing applications, if you have ever written a paper on the computer.

Choose To choose an item from a menu, pull down the menu and, with the button still depressed, move the mouse cursor until the item is highlighted. Then release the mouse button.

Click To click on an item appearing on the screen, move the mouse until the cursor is on top of the item. Then click the button on the mouse.

Clipboard A temporary storage location, used to store text or other information for transferring to another location.

Close box The close box is the small white box in the upper left corner of the window (at the left end of the window's title bar). Clicking in it closes the window.

Command key The command key is the one with the cloverleaf on it. Newer keyboards also have an apple on the command key. To type (say) Command-A, type A while holding down the command key. In other words, use it like you use the shift key when typing upper-case letters.

Deleting files To delete files, you must be at the desktop (see below), and have the disk's contents window open (this is the window showing the Tarski and Exercise Folder icons). You will first need to find the icon for the file you want to delete. If it is in a folder, you will have to open the folder by double-clicking on it. When you find the file, click on it and drag it to the trash can icon. When the trash can turns black, release the button. If you change your mind about deleting the file, retrieve it (before you do anything else) by double-clicking on the trash can and moving the file back to the contents window. If the machine you are working on is running System 7, the trash can is only emptied when you choose **Empty Trash** ... from the **Special** menu. You must do this if you want more space on your disk.

Desktop The desktop is the screen that appears when you are in the Finder. It is the screen with the trash can icon in the lower right corner.

Dialog box A rectangular box that appears on the screen to request or provide information.

Dragging To drag an item from one point on the screen to another, move the mouse until the cursor is on top of the item. Click on the item and, with the button still depressed, move the mouse until the item is where you want it. Then release the button.

Double-click To double-click on an item, move the mouse until the cursor is on top of the item. Then click the button on the mouse twice in quick succession.

File Tarski's World has two kinds of files: sentence files and world files. These are created in the appropriate windows and can be saved on the disk.

Finder The finder is the program that starts up when you first turn on your Macintosh. It displays the trash can icon in the lower right corner and the disk icon(s) in the upper right.

Folder To create a folder, you must be in the Finder. Choose **New Folder** from the Finder's **File** menu, and an empty folder will appear in the contents window. Change the folder's name to the desired name.

Highlight Highlighting is often used to indicate a selected item. Visually, this is indicated by changing the usual colors of the item. For example, black text on a white background becomes white text on a black background when it is selected, and a grey block becomes white when it is selected. This is known as highlighting.

Icon Icons are the graphic representations used by the Macintosh for things like programs, files, disks, and the like.

Initialized disk When you buy a blank disk, it has to be initialized before it can be used. You do this from the Finder (see above). Put the new disk in the spare drive (if your Macintosh has two), and it will ask you if you want to initialize it. Click the appropriate button. It will ask you to name the disk; choose some name other than "Tarski's World," to prevent confusion. If you have only one drive, choose **Eject** from the **File** menu and put the new disk in the drive. Then follow the same procedure.

Insertion point The vertical line that appears in the sentence window when you are writing sentences. It indicates where characters will be inserted when you enter them.

Menu A menu is a list of commands or functions that appears when you click on its title. The menu titles are at the top of the screen, to the right of the apple.

Pull down To pull down a menu, click on the menu title at the top of the screen. Do not release the button until the desired item is chosen. If you decide not to choose an item, simply move the mouse cursor away from the menu.

Scroll Scrolling a window allows you to see portions that are not presently visible. To scroll a window, either click on the up or down arrows in the scroll bar to the right of the window, or drag the scroll box, which appears between the arrows, up or down.

Scroll bar A grey bar with arrows at the top and bottom, and a small square (the scroll box) that moves up and down the bar. The scroll bar is used to move the contents of a window up and down.

Select To select an item, you simply click on it. The item will become highlighted. (In many cases, some item, say the first in a list, will already be highlighted. If that is the one you want to select, then you don't have to do anything.)

Sizing box This is the little box in the lower right corner of the Sentence window. containing two overlapping squares. To change the size of one the window, click in this box and drag.

Title bar The title bar is the bar across the top of a window, usually containing the title of the window or the name of the file displayed in the window.

Window Windows are the rectangular areas on the screen in which all activity must take place.

Appendix C

Summary of Proof Rules

C.1 System $\mathcal{F}$

Propositional Logic

Reflexivity of Identity
(Refl =)

$$\triangleright \quad n = n$$

Indiscernibility of Identicals
(Ind Id)

$$\begin{array}{l|l} & P(n) \\ & \vdots \\ & n = m \\ & \vdots \\ \triangleright & P(m) \end{array}$$

Disjunction Introduction
(∨ Intro)

$$\begin{array}{l|l} & P_i \\ & \vdots \\ \triangleright & P_1 \vee \ldots \vee P_i \vee \ldots \vee P_n \end{array}$$

Disjunction Elimination
(∨ Elim)

$$\begin{array}{l|l} & P_1 \vee \ldots \vee P_n \\ & \vdots \\ & \quad \underline{P_1} \\ & \quad \vdots \\ & \quad S \\ & \Downarrow \\ & \quad \underline{P_n} \\ & \quad \vdots \\ & \quad S \\ & \vdots \\ \triangleright & S \end{array}$$

Reiteration
(Reit)

$$\begin{array}{l|l} & P \\ & \vdots \\ \triangleright & P \end{array}$$

Conjunction Introduction
(∧ Intro)

P_1
⇓
P_n
⋮
▷ $P_1 \wedge \ldots \wedge P_n$

Conjunction Elimination
(∧ Elim)

$P_1 \wedge \ldots \wedge P_i \wedge \ldots \wedge P_n$
⋮
▷ P_i

Negation Introduction
(¬ Intro)

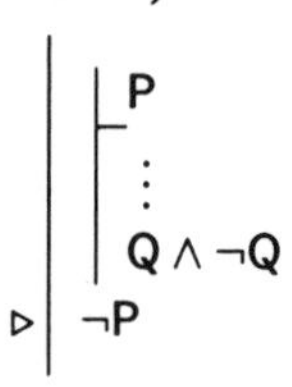

Negation Elimination
(¬ Elim)

$\neg\neg P$
⋮
▷ P

Conditional Introduction
(→ Intro)

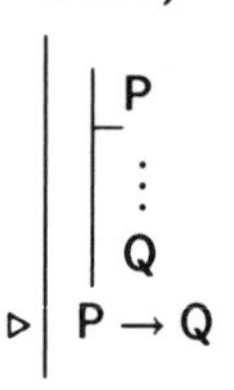

Conditional Elimination
(→ Elim)

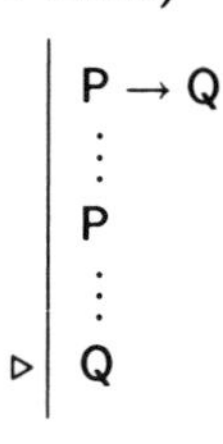

Biconditional Introduction
(↔ Intro)

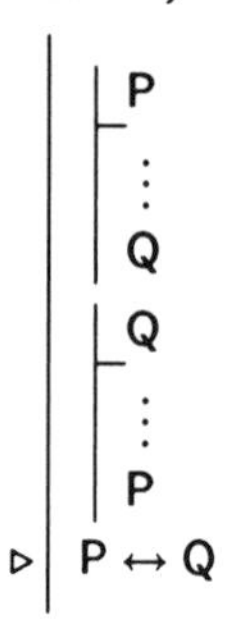

Biconditional Elimination
(↔ Elim)

$P \leftrightarrow Q$ (or $Q \leftrightarrow P$)
⋮
P
⋮
▷ Q

Quantifiers

Universal Introduction
(∀ Intro)

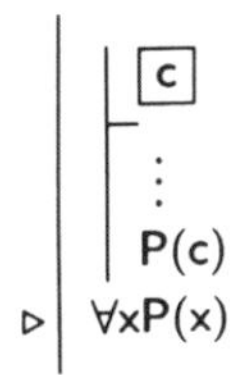

where c does not occur outside the subproof where it is introduced.

Universal Elimination
(∀ Elim)

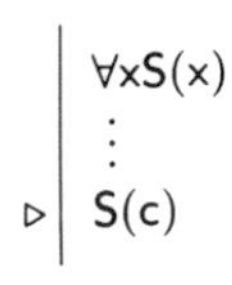

Existential Introduction
(∃ Intro)

$$\mathsf{S(c)}$$
$$\vdots$$
$$\triangleright\ \exists x \mathsf{S}(x)$$

Existential Elimination
(∃ Elim)

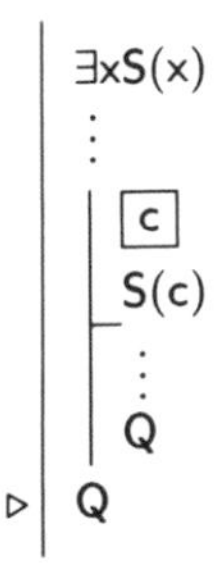

where c does not occur outside the subproof where it is introduced.

C.2 System $\mathcal{F}'$

Propositional Logic

Negation Introduction
(¬ Intro)

Negation Elimination
(¬ Elim)

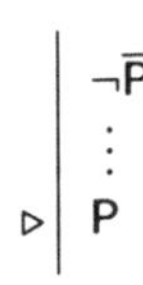

Bottom Introduction ($\perp$ Intro)

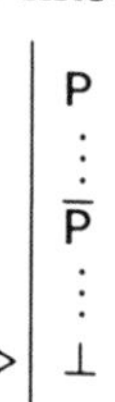

Bottom Elimination ($\perp$ Elim)

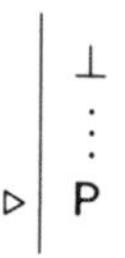

Proof by Contradiction (Contra)

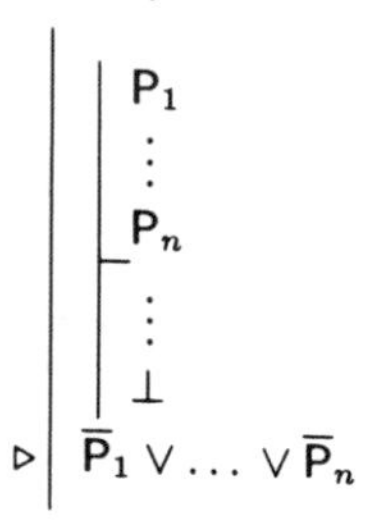

Generalized Substitution (Gen Sub)

$P \leftrightarrow Q$ (or $Q \leftrightarrow P$)
⋮
$S(P)$
⋮
▷ $S(Q)$

Quantifiers

General Conditional Proof (Gen Cond)

$\boxed{c}$
$P(c)$
⋮
$Q(c)$
▷ $\forall x(P(x) \to Q(x))$

where c does not occur outside the subproof where it is introduced.

General Conditional Proof (Gen Cond)

$\boxed{c_1, \ldots, c_n}$
$P(c_1, \ldots, c_n)$
⋮
$Q(c_1, \ldots, c_n)$
▷ $\forall x_1 \ldots \forall x_n(P(x_1, \ldots, x_n) \to Q(x_1, \ldots, x_n))$

where the c_i's are distinct and do not occur outside the subproof where they are introduced.

Universal Introduction
(∀ Intro)

$$\begin{array}{l|l} & \begin{array}{|l} \boxed{c_1, \ldots, c_n} \\ \hline \vdots \\ P(c_1, \ldots, c_n) \end{array} \\ \triangleright & \forall x_1 \ldots \forall x_n P(x_1, \ldots, x_n) \end{array}$$

where the c_i's are distinct and do not occur outside the subproof where they are introduced.

Existential Introduction
(∃ Intro)

$$\begin{array}{l|l} & S(c_1, \ldots, c_n) \\ & \vdots \\ \triangleright & \exists x_1 \ldots \exists x_n S(x_1, \ldots, x_n) \end{array}$$

Uniqueness Introduction
(∃! Intro)

$$\begin{array}{l|l} & P(c) \\ & \vdots \\ & \begin{array}{|l} \boxed{d} \\ P(d) \\ \hline \vdots \\ d = c \end{array} \\ \triangleright & \exists! x P(x) \end{array}$$

where d does not occur outside the subproof where it is introduced.

Universal Elimination
(∀ Elim)

$$\begin{array}{l|l} & \forall x_1 \ldots \forall x_n S(x_1, \ldots, x_n) \\ & \vdots \\ \triangleright & S(c_1, \ldots, c_n) \end{array}$$

Existential Elimination
(∃ Elim)

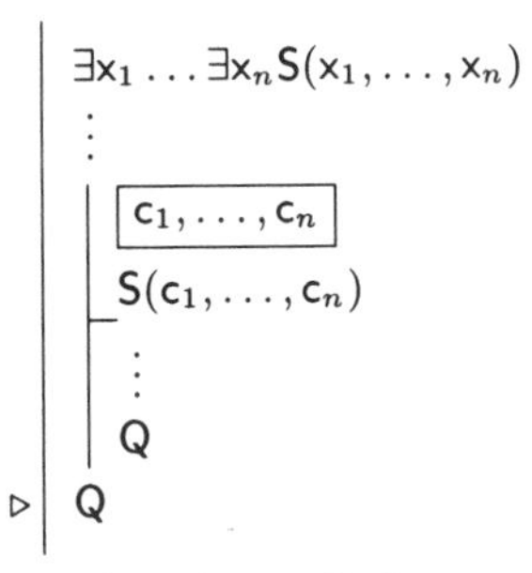

where the c_i's are distinct and do not occur outside the subproof where they are introduced.

Uniqueness Elimination I
(∃! Elim I)

$$\begin{array}{l|l} & \exists! x P(x) \\ & \vdots \\ \triangleright & \exists x P(x) \end{array}$$

Uniqueness Elimination II
(∃! Elim II)

$$\begin{array}{l|l} & \exists! x P(x) \\ & \vdots \\ \triangleright & \forall y \forall z[(P(y) \land P(z)) \to y = z] \end{array}$$

General Index

Tarski's World Index

Exercise Files Index

CSLI Publications

Reports

The following titles have been published in the CSLI Reports series. These reports may be obtained from CSLI Publications, Ventura Hall, Stanford, CA 94305-4115.

Coordination and How to Distinguish Categories Ivan Sag, Gerald Gazdar, Thomas Wasow, and Steven Weisler CSLI–84–3

Belief and Incompleteness Kurt Konolige CSLI–84–4

Equality, Types, Modules and Generics for Logic Programming Joseph Goguen and José Meseguer CSLI–84–5

Lessons from Bolzano Johan van Benthem CSLI–84–6

Self-propagating Search: A Unified Theory of Memory Pentti Kanerva CSLI–84–7

Reflection and Semantics in LISP Brian Cantwell Smith CSLI–84–8

The Implementation of Procedurally Reflective Languages Jim des Rivières and Brian Cantwell Smith CSLI–84–9

Parameterized Programming Joseph Goguen CSLI–84–10

Shifting Situations and Shaken Attitudes Jon Barwise and John Perry CSLI–84–13

Completeness of Many-Sorted Equational Logic Joseph Goguen and José Meseguer CSLI–84–15

Moving the Semantic Fulcrum Terry Winograd CSLI–84–17

On the Mathematical Properties of Linguistic Theories C. Raymond Perrault CSLI–84–18

A Simple and Efficient Implementation of Higher-order Functions in LISP Michael P. Georgeff and Stephen F. Bodnar CSLI–84–19

On the Axiomatization of "if-then-else" Irène Guessarian and José Meseguer CSLI–85–20

The Situation in Logic–II: Conditionals and Conditional Information Jon Barwise CSLI–84–21

Principles of OBJ2 Kokichi Futatsugi, Joseph A. Goguen, Jean-Pierre Jouannaud, and José Meseguer CSLI–85–22

Querying Logical Databases Moshe Vardi CSLI–85–23

Computationally Relevant Properties of Natural Languages and Their Grammar Gerald Gazdar and Geoff Pullum CSLI–85–24

An Internal Semantics for Modal Logic: Preliminary Report Ronald Fagin and Moshe Vardi CSLI–85–25

The Situation in Logic–III: Situations, Sets and the Axiom of Foundation Jon Barwise CSLI–85–26

Semantic Automata Johan van Benthem CSLI–85–27

Restrictive and Non-Restrictive Modification Peter Sells CSLI–85–28

Institutions: Abstract Model Theory for Computer Science J. A. Goguen and R. M. Burstall CSLI–85–30

A Formal Theory of Knowledge and Action Robert C. Moore CSLI–85–31

Finite State Morphology: A Review of Koskenniemi (1983) Gerald Gazdar CSLI–85–32

The Role of Logic in Artificial Intelligence Robert C. Moore CSLI–85–33

Applicability of Indexed Grammars to Natural Languages Gerald Gazdar CSLI–85–34

Commonsense Summer: Final Report Jerry R. Hobbs, et al CSLI–85–35

Limits of Correctness in Computers Brian Cantwell Smith CSLI–85–36

The Coherence of Incoherent Discourse Jerry R. Hobbs and Michael H. Agar CSLI–85–38

A Complete, Type-free "Second-order" Logic and Its Philosophical Foundations Christopher Menzel CSLI–86–40

Possible-world Semantics for Autoepistemic Logic Robert C. Moore CSLI–85–41

Deduction with Many-Sorted Rewrite José Meseguer and Joseph A. Goguen CSLI–85–42

On Some Formal Properties of Metarules Hans Uszkoreit and Stanley Peters CSLI–85–43

Language, Mind, and Information John Perry CSLI–85–44

Constraints on Order Hans Uszkoreit CSLI–86–46

Linear Precedence in Discontinuous Constituents: Complex Fronting in German Hans Uszkoreit CSLI–86–47

A Compilation of Papers on Unification-Based Grammar Formalisms, Parts I and II Stuart M. Shieber, Fernando C.N. Pereira, Lauri Karttunen, and Martin Kay CSLI–86–48

Noun-Phrase Interpretation Mats Rooth CSLI–86–51

Noun Phrases, Generalized Quantifiers and Anaphora Jon Barwise CSLI–86–52

Circumstantial Attitudes and Benevolent Cognition John Perry CSLI–86–53

A Study in the Foundations of Programming Methodology: Specifications, Institutions, Charters and Parchments Joseph A. Goguen and R. M. Burstall CSLI–86–54

Intentionality, Information, and Matter Ivan Blair CSLI–86–56

Computer Aids for Comparative Dictionaries Mark Johnson CSLI–86–58

A Sheaf-Theoretic Model of Concurrency Luís F. Monteiro and Fernando C. N. Pereira CSLI–86–62

Tarski on Truth and Logical Consequence John Etchemendy CSLI–86–64

Categorial Unification Grammars Hans Uszkoreit CSLI–86–66

Generalized Quantifiers and Plurals Godehard Link CSLI–86–67

Radical Lexicalism Lauri Karttunen CSLI–86–68

What is Intention? Michael E. Bratman CSLI–86–69

The Correspondence Continuum Brian Cantwell Smith CSLI–87–71

The Role of Propositional Objects of Belief in Action David J. Israel CSLI–87–72

Two Replies Jon Barwise CSLI–87–74

Semantics of Clocks Brian Cantwell Smith CSLI–87–75

The Parts of Perception Alexander Pentland CSLI–87–77

The Situated Processing of Situated Language Susan Stucky CSLI–87–80

Muir: A Tool for Language Design Terry Winograd CSLI–87–81

Final Algebras, Cosemicomputable Algebras, and Degrees of Unsolvability Lawrence S. Moss, José Meseguer, and Joseph A. Goguen CSLI–87–82

The Synthesis of Digital Machines with Provable Epistemic Properties Stanley J. Rosenschein and Leslie Pack Kaelbling CSLI–87–83

An Architecture for Intelligent Reactive Systems Leslie Pack Kaelbling CSLI–87–85

Modular Algebraic Specification of Some Basic Geometrical Constructions Joseph A. Goguen CSLI–87–87

Persistence, Intention and Commitment Phil Cohen and Hector Levesque CSLI–87–88

Rational Interaction as the Basis for Communication Phil Cohen and Hector Levesque CSLI–87–89

Models and Equality for Logical Programming Joseph A. Goguen and José Meseguer CSLI–87–91

Order-Sorted Algebra Solves the Constructor-Selector, Mulitple Representation and Coercion Problems Joseph A. Goguen and José Meseguer CSLI–87–92

Extensions and Foundations for Object-Oriented Programming Joseph A. Goguen and José Meseguer CSLI–87–93

L3 Reference Manual: Version 2.19 William Poser CSLI–87–94

Change, Process and Events Carol E. Cleland CSLI–88–95

One, None, a Hundred Thousand Specification Languages Joseph A. Goguen CSLI–87–96

Constituent Coordination in HPSG Derek Proudian and David Goddeau CSLI–87–97

A Language/Action Perspective on the Design of Cooperative Work Terry Winograd CSLI–87–98

Implicature and Definite Reference Jerry R. Hobbs CSLI–87–99

Situation Semantics and Semantic Interpretation in Constraint-based Grammars Per-Kristian Halvorsen CSLI–87–101

Category Structures Gerald Gazdar, Geoffrey K. Pullum, Robert Carpenter, Ewan Klein, Thomas E. Hukari, Robert D. Levine CSLI–87–102

Cognitive Theories of Emotion Ronald Alan Nash CSLI–87–103

Toward an Architecture for Resource-bounded Agents Martha E. Pollack, David J. Israel, and Michael E. Bratman CSLI–87–104

On the Relation Between Default and Autoepistemic Logic Kurt Konolige CSLI–87–105

Three Responses to Situation Theory Terry Winograd CSLI–87–106

Subjects and Complements in HPSG Robert Borsley CSLI–87–107

Tools for Morphological Analysis Mary Dalrymple, Ronald M. Kaplan, Lauri Karttunen, Kimmo Koskenniemi, Sami Shaio, Michael Wescoat CSLI–87–108

Fourth Year Report of the Situated Language Research Program CSLI–87–111

Events and "Logical Form" Stephen Neale CSLI–88–113

Backwards Anaphora and Discourse Structure: Some Considerations Peter Sells CSLI–87–114

Toward a Linking Theory of Relation Changing Rules in LFG Lori Levin CSLI–87–115

Fuzzy Logic L. A. Zadeh CSLI–88–116

Dispositional Logic and Commonsense Reasoning L. A. Zadeh CSLI–88–117

Intention and Personal Policies Michael Bratman CSLI–88–118

Unification and Agreement Michael Barlow CSLI–88–120

Extended Categorial Grammar Suson Yoo and Kiyong Lee CSLI–88–121

Unaccusative Verbs in Dutch and the Syntax-Semantics Interface Annie Zaenen CSLI–88–123

Types and Tokens in Linguistics Sylvain Bromberger CSLI–88–125

Determination, Uniformity, and Relevance: Normative Criteria for Generalization and Reasoning by Analogy Todd Davies CSLI–88–126

Modal Subordination and Pronominal Anaphora in Discourse Craige Roberts CSLI–88–127

The Prince and the Phone Booth: Reporting Puzzling Beliefs Mark Crimmins and John Perry CSLI–88–128

Set Values for Unification-Based Grammar Formalisms and Logic Programming William Rounds CSLI–88–129

Fifth Year Report of the Situated Language Research Program CSLI–88–130

Locative Inversion in Chicheŵa: A Case Study of Factorization in Grammar Joan Bresnan and Jonni M. Kanerva CSLI–88–131

An Information-Based Theory of Agreement Carl Pollard and Ivan A. Sag CSLI–88–132

Relating Models of Polymorphism José Meseguer CSLI–88–133

Psychology, Semantics, and Mental Events under Descriptions Peter Ludlow CSLI–89–135

Mathematical Proofs of Computer System Correctness Jon Barwise CSLI–89–136

The X-bar Theory of Phrase Structure András Kornai and Geoffrey K. Pullum CSLI–89–137

Discourse Structure and Performance Efficiency in Interactive and Noninteractive Spoken Modalities Sharon L. Oviatt and Philip R. Cohen CSLI–90–138

The Contributing Influence of Speech and Interaction on Some Aspects of Human Discourse Sharon L. Oviatt and Philip R. Cohen CSLI–90–139

The Connectionist Construction of Concepts Adrian Cussins CSLI–90–140

Sixth Year Report CSLI–90–141

Categorical Grammar Meets Unification Johan van Benthem CSLI–90–142

Point of View Edit Doron CSLI–90–143

Modal Logic as a Theory of Information Johan van Benthem CSLI–90–144

What Is Information? David Israel and John Perry CSLI–91–145

Fodor and Psychological Explanations John Perry and David Israel CSLI–91–146

Decision Problems for Propositional Linear Logic Patrick Lincoln, John Mitchell, André Scedrov, and Natarajan Shankar CSLI–91–147

Annual Report 1989-90 CSLI–91–148

Overloading Intentions for Efficient Practical Reasoning Martha E. Pollack CSLI–91–149

Introduction to the Project on People, Computers, and Design Terry Winograd CSLI–91–150

Ecological Psychology and Dewey's Theory of Perception Tom Burke CSLI–91–151

The Language/Action Approach to the Design of Computer-Support for Cooperative Work Finn Kensing and Terry Winograd CSLI–91–152

The Absorption Principle and E-Type Anaphora Jean Mark Gawron, John Nerbonne, and Stanley Peters CSLI–91–153

Ellipsis and Higher-Order Unification Mary Dalrymple, Stuart M. Shieber, and Fernando C. N. Pereira CSLI–91–154

Sheaf Semantics for Concurrent Interacting Objects Joseph A. Goguen CSLI–91–155

Communication and Strategic Inference Prashant Parikh CSLI–91–156

Shared Cooperative Activity Michael E. Bratman CSLI–91–157

Practical Reasoning and Acceptance in a Context Michael E. Bratman CSLI–91–158

Planning and the Stability of Intention Michael E. Bratman CSLI–91–159

Logic and the Flow of Information Johan van Benthem CSLI–91–160

Learning HCI Design: Mentoring Project Groups in a Course on Human-Computer Interaction Brad Hartfield, Terry Winograd, and John Bennett CSLI–91–161

How to Read Winograd's & Flores's Understanding Computers and Cognition Hugh McGuire CSLI–92–162

In Support of a Semantic Account of Resultatives Adele E. Goldberg CSLI–92–163

Augmenting Informativeness and Learnability of Items in Large Computer Networks Clarisse S. de Souza CSLI–92–164

Computers, Ethics, and Social Responsibility Terry Winograd CSLI–92–165

A Semiotic Approach to User Interface Language Design Clarisse S. de Souza CSLI–92–166

1991 Annual Report CSLI–92–167

Romance is so complex Christopher Manning CSLI–92–168

Types, Tokens and Templates David M. Levy and Kenneth R. Olson CSLI–92–169

A System of Dynamic Modal Logic Maarten de Rijke CSLI–92–170

Situation Theory and the Design of Interactive Information Systems Keith Devlin CSLI–92–171

Logic as Programming Johan van Benthem CSLI–92–172

An Architecture for Tuning Rules and Cases Yoshio Nakatami and David Israel CSLI–92–173

The Linguistic Information in Dynamic Discourse Megumi Kameyama CSLI–92–174

Epsilon Substitution Method for Elementary Analysis Grigori Mints and Sergei Tupailo CSLI–93–175

Information Spreading and Levels of Representation in LFG Avery D. Andrews and Christopher D. Manning CSLI–93–176

1992 Annual Report CSLI–93–177

Lecture Notes

The titles in this series are distributed by the University of Chicago Press and may be purchased in academic or university bookstores or ordered directly from the distributor at 11030 South Langley Avenue, Chicago, IL 60628 (USA) or by phone 1-800-621-2736, (312)568-1550.

A Manual of Intensional Logic. Johan van Benthem, second edition, revised and expanded. Lecture Notes No. 1. 0-937073-29-6 (paper), 0-937073-30-X (cloth)

Emotion and Focus. Helen Fay Nissenbaum. Lecture Notes No. 2. 0-937073-20-2 (paper)

Lectures on Contemporary Syntactic Theories. Peter Sells. Lecture Notes No. 3. 0-937073-14-8 (paper), 0-937073-13-X (cloth)

An Introduction to Unification-Based Approaches to Grammar. Stuart M. Shieber. Lecture Notes No. 4. 0-937073-00-8 (paper), 0-937073-01-6 (cloth)

The Semantics of Destructive Lisp. Ian A. Mason. Lecture Notes No. 5. 0-937073-06-7 (paper), 0-937073-05-9 (cloth)

An Essay on Facts. Ken Olson. Lecture Notes No. 6. 0-937073-08-3 (paper), 0-937073-05-9 (cloth)

Logics of Time and Computation. Robert Goldblatt, second edition, revised and expanded. Lecture Notes No. 7. 0-937073-94-6 (paper), 0-937073-93-8 (cloth)

Word Order and Constituent Structure in German. Hans Uszkoreit. Lecture Notes No. 8. 0-937073-10-5 (paper), 0-937073-09-1 (cloth)

Color and Color Perception: A Study in Anthropocentric Realism. David Russel Hilbert. Lecture Notes No. 9. 0-937073-16-4 (paper), 0-937073-15-6 (cloth)

Prolog and Natural-Language Analysis. Fernando C. N. Pereira and Stuart M. Shieber. Lecture Notes No. 10. 0-937073-18-0 (paper), 0-937073-17-2 (cloth)

Working Papers in Grammatical Theory and Discourse Structure: Interactions of Morphology, Syntax, and Discourse. M. Iida, S. Wechsler, and D. Zec (Eds.) with an Introduction by Joan Bresnan. Lecture Notes No. 11. 0-937073-04-0 (paper), 0-937073-25-3 (cloth)

Natural Language Processing in the 1980s: A Bibliography. Gerald Gazdar, Alex Franz, Karen Osborne, and Roger Evans. Lecture Notes No. 12. 0-937073-28-8 (paper), 0-937073-26-1 (cloth)

Information-Based Syntax and Semantics. Carl Pollard and Ivan Sag. Lecture Notes No. 13. 0-937073-24-5 (paper), 0-937073-23-7 (cloth)

Non-Well-Founded Sets. Peter Aczel. Lecture Notes No. 14. 0-937073-22-9 (paper), 0-937073-21-0 (cloth)

Partiality, Truth and Persistence. Tore Langholm. Lecture Notes No. 15. 0-937073-34-2 (paper), 0-937073-35-0 (cloth)

Attribute-Value Logic and the Theory of Grammar. Mark Johnson. Lecture Notes No. 16. 0-937073-36-9 (paper), 0-937073-37-7 (cloth)

The Situation in Logic. Jon Barwise. Lecture Notes No. 17. 0-937073-32-6 (paper), 0-937073-33-4 (cloth)

The Linguistics of Punctuation. Geoff Nunberg. Lecture Notes No. 18. 0-937073-46-6 (paper), 0-937073-47-4 (cloth)

Anaphora and Quantification in Situation Semantics. Jean Mark Gawron and Stanley Peters. Lecture Notes No. 19. 0-937073-48-4 (paper), 0-937073-49-0 (cloth)

Propositional Attitudes: The Role of Content in Logic, Language, and Mind. C. Anthony Anderson and Joseph Owens. Lecture Notes No. 20. 0-937073-50-4 (paper), 0-937073-51-2 (cloth)

Literature and Cognition. Jerry R. Hobbs. Lecture Notes No. 21. 0-937073-52-0 (paper), 0-937073-53-9 (cloth)

Situation Theory and Its Applications, Vol. 1. Robin Cooper, Kuniaki Mukai, and John Perry (Eds.). Lecture Notes No. 22. 0-937073-54-7 (paper), 0-937073-55-5 (cloth)

The Language of First-Order Logic (including the Macintosh program, Tarski's World). Jon Barwise and John Etchemendy, second edition, revised and expanded. Lecture Notes No. 23. 0-937073-74-1 (paper)

Lexical Matters. Ivan A. Sag and Anna Szabolcsi, editors. Lecture Notes No. 24. 0-937073-66-0 (paper), 0-937073-65-2 (cloth)

Tarski's World. Jon Barwise and John Etchemendy. Lecture Notes No. 25. 0-937073-67-9 (paper)

Situation Theory and Its Applications, Vol. 2. Jon Barwise, J. Mark Gawron, Gordon Plotkin, Syun Tutiya, editors. Lecture Notes No. 26. 0-937073-70-9 (paper), 0-937073-71-7 (cloth)

Literate Programming. Donald E. Knuth. Lecture Notes No. 27. 0-937073-80-6 (paper), 0-937073-81-4 (cloth)

Normalization, Cut-Elimination and the Theory of Proofs. A. M. Ungar. Lecture Notes No. 28. 0-937073-82-2 (paper), 0-937073-83-0 (cloth)

Lectures on Linear Logic. A. S. Troelstra. Lecture Notes No. 29. 0-937073-77-6 (paper), 0-937073-78-4 (cloth)

A Short Introduction to Modal Logic. Grigori Mints. Lecture Notes No. 30. 0-937073-75-X (paper), 0-937073-76-8 (cloth)

Linguistic Individuals. Almerindo E. Ojeda. Lecture Notes No. 31. 0-937073-84-9 (paper), 0-937073-85-7 (cloth)

Computer Models of American Speech. M. Margaret Withgott and Francine R. Chen. Lecture Notes No. 32. 0-937073-98-9 (paper), 0-937073-97-0 (cloth)

Verbmobil: A Translation System for Face-to-Face Dialog. Martin Kay, Mark Gawron, and Peter Norvig. Lecture Notes No. 33. 0-937073-95-4 (paper), 0-937073-96-2 (cloth)

The Language of First-Order Logic (including the Windows program, Tarski's World). Jon Barwise and John Etchemendy, third edition, revised and expanded. Lecture Notes No. 34. 0-937073-90-3 (paper)

Turing's World. Jon Barwise and John Etchemendy. Lecture Notes No. 35. 1-881526-10-0 (paper)

Syntactic Constraints on Anaphoric Binding. Mary Dalrymple. Lecture Notes No. 36. 1-881526-06-2 (paper), 1-881526-07-0 (cloth)

Situation Theory and Its Applications, Vol. 3. Peter Aczel, David Israel, Yasuhiro Katagiri, and Stanley Peters, editors. Lecture Notes No. 37. 1-881526-08-9 (paper), 1-881526-09-7 (cloth)

Theoretical Aspects of Bantu Grammar. Mchombo, editor. Lecture Notes No. 38. 0-937073-72-5 (paper), 0-937073-73-3 (cloth)

Logic and Representation. Robert C. Moore. Lecture Notes No. 39. 1-881526-15-1 (paper), 1-881526-16-X (cloth)

Meanings of Words and Contextual Determination of Interpretation. Paul Kay. Lecture Notes No. 40. 1-881526-17-8 (paper), 1-881526-18-6 (cloth)

Language and Learning for Robots. Colleen Crangle and Patrick Suppes. Lecture Notes No. 41. 1-881526-19-4 (paper), 1-881526-20-8 (cloth)

Hyperproof. Jon Barwise and John Etchemendy. Lecture Notes No. 42. 1-881526-11-9 (paper)

Mathematics of Modality. Robert Goldblatt. Lecture Notes No. 43. 1-881526-23-2 (paper), 1-881526-24-0 (cloth)

Feature Logics, Infinitary Descriptions, and Grammar. Bill Keller. Lecture Notes No. 44. 1-881526-25-9 (paper), 1-881526-26-7 (cloth)

Other CSLI Titles Distributed by UCP

Agreement in Natural Language: Approaches, Theories, Descriptions. Michael Barlow and Charles A. Ferguson, editors. 0-937073-02-4 (cloth)

Papers from the Second International Workshop on Japanese Syntax. William J. Poser, editor. 0-937073-38-5 (paper), 0-937073-39-3 (cloth)

The Proceedings of the Seventh West Coast Conference on Formal Linguistics (WCCFL 7). 0-937073-40-7 (paper)

The Proceedings of the Eighth West Coast Conference on Formal Linguistics (WCCFL 8). 0-937073-45-8 (paper)

The Phonology-Syntax Connection. Sharon Inkelas and Draga Zec (Eds.) (co-published with The University of Chicago Press). 0-226-38100-5 (paper), 0-226-38101-3 (cloth)

The Proceedings of the Ninth West Coast Conference on Formal Linguistics (WCCFL 9). 0-937073-64-4 (paper)

Japanese/Korean Linguistics. Hajime Hoji, editor. 0-937073-57-1 (paper), 0-937073-56-3 (cloth)

Experiencer Subjects in South Asian Languages. Manindra K. Verma and K. P. Mohanan, editors. 0-937073-60-1 (paper), 0-937073-61-X (cloth)

Grammatical Relations: A Cross-Theoretical Perspective. Katarzyna Dziwirek, Patrick Farrell, Errapel Mejías Bikandi, editors. 0-937073-63-6 (paper), 0-937073-62-8 (cloth)

The Proceedings of the Tenth West Coast Conference on Formal Linguistics (WCCFL 10). 0-937073-79-2 (paper)

On What We Know We Don't Know. Sylvain Bromberger. 0-226-075400 (paper), (cloth)

The Proceedings of the Twenty-fourth Annual Child Language Research Forum. Eve V. Clark, editor. 1-881526-05-4 (paper), 1-881526-04-6 (cloth)

Japanese/Korean Linguistics, Vol. 2. Patricia M. Clancy, editor. 1-881526-13-5 (paper), 1-881526-14-3 (cloth)

Arenas of Language Use. Herbert H. Clark. 0-226-10782-5 (paper), (cloth)

Japanese/Korean Linguistics, Vol. 3. Sonja Choi, editor. 1-881526-21-6 (paper), 1-881526-22-4 (cloth)

The Proceedings of the Eleventh West Coast Conference on Formal Linguistics (WCCFL 11). 1-881526-12-7 (paper)

Books Distributed by CSLI

The Proceedings of the Third West Coast Conference on Formal Linguistics (WCCFL 3). 0-937073-44-X (paper)

The Proceedings of the Fourth West Coast Conference on Formal Linguistics (WCCFL 4). 0-937073-43-1 (paper)

The Proceedings of the Fifth West Coast Conference on Formal Linguistics (WCCFL 5). 0-937073-42-3 (paper)

The Proceedings of the Sixth West Coast Conference on Formal Linguistics (WCCFL 6). 0-937073-31-8 (paper)

Hausar Yau Da Kullum: Intermediate and Advanced Lessons in Hausa Language and Culture. William R. Leben, Ahmadu Bello Zaria, Shekarau B. Maikafi, and Lawan Danladi Yalwa. 0-937073-68-7 (paper)

Hausar Yau Da Kullum Workbook. William R. Leben, Ahmadu Bello Zaria, Shekarau B. Maikafi, and Lawan Danladi Yalwa. 0-93703-69-5 (paper)

Ordering Titles Distributed by CSLI

Titles distributed by CSLI may be ordered directly from CSLI Publications, Ventura Hall, Stanford, CA 94305-4115 or by phone (415)723-1712, (415)723-1839. Orders can also be placed by FAX (415)723-0758 or e-mail (pubs@csli.stanford.edu).

All orders must be prepaid by check or Visa or MasterCard (include card name, number, and expiration date). California residents add 8.25% sales tax. For shipping and handling, add $2.50 for first book and $0.75 for each additional book; $1.75 for first report and $0.25 for each additional report.

For overseas shipping, add $4.50 for first book and $2.25 for each additional book; $2.25 for first report and $0.75 for each additional report. All payments must be made in U.S. currency.

READ THIS INFORMATION BEFORE OPENING THE POUCH CONTAINING THE DISKETTE

Please note that this book and the diskette accompanying it may not be returned after the pouch containing the diskette has been unsealed. The diskette, Tarski's World 4.0, is protected by copyright and may not be duplicated, except for purposes of archival back up. All inquiries concerning the diskette should be directed to the Publications Office, CSLI, Ventura Hall, Stanford University, Stanford, CA 94305; phone: (415)723–1712 or (415)723–1839. The program has been duplicated on the finest quality diskettes, verified before shipment. If a diskette is damaged, please send it to the above address for replacement. Please note also that the distributor of this book and program, the University of Chicago Press, is not equipped to handle inquiries about this package, nor to replace defective diskettes.

A NOTE TO BOOKSTORES

The publisher recommends that bookstores not sell used copies of this text-software package. Bookstores selling used copies must be prepared to warrant that 1) the original owner retains no copies of the software and 2) the disk contains all of the exercise files in their original form and contains no solution files created by the initial owner.